Novel Nanocrystalline Alloys
and Magnetic Nanomaterials

T0143563

Series in Materials Science and Engineering

Series Editors: **B Cantor**, University of York, UK
M J Goringe, School of Mechanical and Materials Engineering, University of Surrey, UK
E Ma, Department of Materials Science and Engineering, Johns Hopkins University, USA

Other titles in the series

Microelectronic Materials
C R M Grovenor

Aerospace Materials
B Cantor, H Assender and P Grant (eds)

Fundamentals of Ceramics
M Barsoum

Solidification and Casting
B Cantor and K O'Reilly (eds)

Topics in the Theory of Solid Materials
J M Vail

Physical Methods for Materials Characterization: Second Edition
P E J Flewitt and R K Wild

Metal and Ceramic Matrix Composites
B Cantor, F P E Dunne and I C Stone (eds)

High Pressure Surface Science
Y Gogotsi and V Domnich (eds)

High-K Gate Dielectrics
M Houssa (ed)

Computer Modelling of Heat and Fluid Flow in Materials Processing
C-P Hong

Forthcoming titles in the series

Fundamentals of Fibre Reinforced Composite Materials
A R Bunsell and J Renard

Series in Materials Science and Engineering

Novel Nanocrystalline Alloys and Magnetic Nanomaterials

An Oxford–Kobe Materials Text

Edited by

Brian Cantor
University of York, UK

CRC Press
Taylor & Francis Group
Boca Raton London New York

CRC Press is an imprint of the
Taylor & Francis Group, an **informa** business

CRC Press
Taylor & Francis Group
6000 Broken Sound Parkway NW, Suite 300
Boca Raton, FL 33487-2742

First issued in paperback 2019

ISBN-13: 978-0-7503-1002-4 (hbk)
ISBN-13: 978-0-367-39364-9 (pbk)

British Library Cataloguing-in-Publication Data

A catalogue record for this book is available from the British Library.

Library of Congress Cataloging-in-Publication Data are available

Typeset by Academic + Technical, Bristol
Index by Indexing Specialists (UK) Ltd, Hove, East Sussex

**Visit the Taylor & Francis Web site at
http://www.taylorandfrancis.com**

**and the CRC Press Web site at
http://www.crcpress.com**

Contents

Preface

This book is a text on novel nanocrystalline alloys and magnetic nano-materials arising out of presentations given at the fourth Oxford–Kobe Materials Seminar, held at the Kobe Institute on 11–14 September 2001.

The Kobe Institute is an independent non profit-making organization. It was established by donations from Kobe City, Hyogo Prefecture and more than 100 companies all over Japan. It is based in Kobe City, Japan, and is operated in collaboration with St Catherine's College, Oxford University, UK. The Chairman of the Kobe Institute Committee in the UK is Roger Ainsworth, Master of St Catherine's College; the Director of the Kobe Institute Board is Dr Yasutomi Nishizuka; the Academic Director is Dr Helen Mardon, Oxford University; and the Bursar is Dr Kaizaburo Saito. The Kobe Institute was established with the objectives of promoting the pursuit of education and research that furthers mutual understanding between Japan and other nations, and to contribute to collaboration and exchange between academics and industrial partners.

The Oxford–Kobe Seminars are research workshops which aim to promote international academic exchanges between the UK/Europe and Japan. A key feature of the seminars is to provide a world-class forum focused on strengthening connections between academics and industry in both Japan and the UK/Europe, and fostering collaborative research on timely problems of mutual interest.

The fourth Oxford–Kobe Materials Seminar was on nanomaterials, concentrating on developments in science and technology over the next ten years. The co-chairs of the Seminar were Professor Akihisa Inoue of Tohoku University, Dr Yoshihito Yoshizawa of Hitachi Metals, Professor Brian Cantor of York University, Dr Paul Warren of Oxford University and Dr Kaizaburo Saito of the Kobe Institute. The Seminar Coordinator was Ms Pippa Gordon of Oxford University. The Seminar was sponsored by the Kobe Institute, St Catherine's College and the Oxford Centre for Advanced Materials and Composites. Following the Seminar itself, all of the speakers prepared extended manuscripts in order to compile a text suitable for graduates and for researchers entering the field. The contributions are compiled into three sections: nanocrystalline alloys, novel nanomaterials, and magnetic nanomaterials.

The first, second and third Oxford–Kobe Materials Seminars were on aerospace materials in September 1998, solidification and casting in September 1999, and metal and ceramic composites in September 2000. The corresponding texts have already been published in the Institute of Physics Publishing Series in Materials Science and Engineering. The fifth, sixth and seventh Oxford–Kobe Materials Seminars were on automotive materials in September 2002, magnetic materials in September 2003 and spintronic materials in September 2004 respectively. The corresponding texts are currently in press in the Institute of Physics Publishing Series in Materials Science and Engineering.

Acknowledgments

Brian Cantor

The editor would like to thank the following: the Oxford–Kobe Institute Committee and St Catherine's College, Oxford University, for agreeing to support the Oxford–Kobe Materials Seminar on Nanomaterials; Sir Peter Williams, Dr Yoshihito Yoshizawa, Dr Paul Warren, Dr Helen Mardon and Kaizaburo Saito for help in organizing the Seminar; and Ms Pippa Gordon, Ms Sarah French and Ms Linda Barton for help with preparing the manuscripts.

Individual authors would like to make additional acknowledgments as follows.

Yoshihiko Hirotsu, Tadakatsu Okhuba and Mitsuhide Matsushita

These investigations were partly supported by a Grant-in-Aid for Scientific Research on Priority Areas of the Ministry of Education, Science and Culture, Japan, and also supported by Special Coordination Funds for Promoting Science and Technology on the Nanohetero Metallic Materials programme from the Science and Technology Agency.

Kazuhiko Kita, Hiroyuki Sasaki, Junichiro Nagahora and Akihisa Inoue

This work was supported by the New Energy and Industrial Technology Development Organization (NEDO) as a part of the Super Metal Technology project.

Tohru Yamasaki

The author is deeply grateful to Professors Ogino and Mochizuki from Himeji Institute of Technology, Japan, for their useful discussions. The author also gratefully acknowledges financial support from the Kawanishi-Memorial Shinmaywa Scientific Foundation Japan (2001) and the Grant-in-Aid from the Japanese Ministry of Education, Culture, Sports, Science and Technology (2001).

Do Hyang Kim

This research was supported by the Creative Research Initiative of the Korean Ministry of Science and Technology.

Eiichiro Matsubara and Takahiro Nakamura

Anomalous x-ray scattering measurements were obtained using synchrotron radiation at the Photon Factory of the Institute of Materials Structure Science (IMSS) under proposal No. 2000G239. The work is partly supported by Special Coordination Funds for Promoting Science and Technology on Nanohetero Metallic Materials from the Science and Technology Agency. The work is also financially supported by a Grant-in-Aid for Scientific Research on Priority Areas, (B)(2) (No. 12130201) from the Ministry of Education, Culture, Sports, Science and Technology of Japan.

Minoro Umemoto

This work was partly supported by the Ferrous Super Metal Consortium of Japan under the auspices of NEDO and the Grant-in-Aid by the Japan Society for the Promotion of Science. The author thanks Dr K Tsuchiya, Dr Z G Liu, Dr Y Xu, Ms J Yin and Mr Suzuki for their involvement

Terence Langdon

Cooperation in preparing this chapter was made possible through an appointment as Visiting Professor in the Department of Materials Science and Engineering at Kyushu University, Fukuoka, Japan, with support from the Japan Society for the Promotion of Science.

De Liang Zhang

Research on metal–ceramic nanocomposites at University of Waikato is funded by the Foundation for Research, Science and Technology, New Zealand, through the New Economy Research Fund (NERF) scheme. The author would also like to thank Mr Jing Liang for his assistance in preparing the manuscript.

K Takanashi, S Mitani, K Yakushiji and H Fujimori

The authors are grateful to Dr H Imamura, Dr S Takahashi, Dr J Martinek, and Professor S Maekawa, Tohoku University, for their theoretical support and useful discussion. Thanks are also given to Dr K Hono, National Institute for Materials Science, for TEM observation, and to Professor J Q Xiao, University of Delaware, for comments on a draft of the manuscript. This work was supported by a JSPS Research Project for the Future

Program (JSPS-RFTF 96P00106) and by CREST of Japan Science and Technology Corporation.

Akihiro Makino

This work was partly supported by Special Coordination Funds for Promoting Science and Technology on Nanohetero Metallic Materials from the Ministry of Education, Culture, Sports, Science and Technology.

Yoshihito Yoshizawa

The author is grateful to Dr K Hono, Dr D H Ping, and Dr M Ohnuma of the National Institute for Materials Science for valuable discussions and nanostructure analysis.

Rainer Hilzinger

The author gratefully acknowledges many valuable suggestions and discussions with R Wengerter, G Herzer and J Petzold.

Satoshi Hirosawa

Collaboration with the author's colleagues, Hirokazu Kanekiyo, Yasutaka Shigemoto, Kaichi Murakami, Toshio Miyoshi and Yusuke Shioya is gratefully acknowledged. Construction of the technique to measure cooling behaviour during rapid solidification was achieved under the Nanohetero Metallic Materials programme as part of the Special Coordination Funds for Promoting Science and Technology from National Institute for Materials Science.

SECTION 1

NANOCRYSTALLINE ALLOYS

A wide variety of nanocrystalline metallic alloys have been developed in recent years, manufactured by rapid cooling from the liquid, condensation from the vapour, electrodeposition and heat treatment of an amorphous precursor. These nanocrystalline alloys have produced considerable scientific interest in trying to understand the relationships between processing, structure and properties, and considerable technological interest in trying to make use of the resulting wide variety of novel mechanical, magnetic and other properties in industrial products.

This section is concerned with nanocrystalline metallic alloys. Chapters 1 and 2 describe the fundamental thermodynamics and structure of nanocrystalline alloys. Chapters 3–5 discuss the manufacture, structure and properties of a variety of different nanocrystalline aluminium alloys; chapters 6–8 discuss the manufacture, structure and properties of a variety of different nanocrystalline nickel, copper, titanium and zirconium alloys; and chapter 9 discusses the manufacture, structure and properties of metallic nanocomposites.

Chapter 1

Thermodynamics of nanocrystalline materials

Livio Battezzati

Introduction

Materials termed nanocrystalline may display very different structures. A first type is entirely made of small crystals produced by techniques for atom condensation and eventual compaction. A second type consists of ensembles of particles deposited on top of a substrate to form a film or a coating. The material may be nanostructured because of the thickness of the film and/or because the film contains fine particles. A third type is represented by bulk solids processed by thermal and/or mechanical means so that crystal sizes are in the nanometre range [1]. There is no unique thermodynamic description for all categories. It is usually stated that materials composed fully or partially of nanocrystals are not in stable equilibrium since they retain a large amount of excess free energy [2]. This is the more widespread case which will be dealt with in this chapter. However, there are prominent examples of nano-structured systems which are described as being in equilibrium since they form from spontaneous organization of molecules.

For all types of nanostructures an essential feature is the structure and stability of the interfaces. Here again there are several examples such as grain boundaries, interphase boundaries, triple junctions, internal channels and cages and external interface [3]. To describe their thermodynamics it is necessary to consider structure, curvature and segregation/adsorption effects.

This chapter describes the hierarchy in stability of nanostructures and the thermodynamics of single phase materials (driving forces for recrystallization, lowering of the melting point). Then, multicomponent materials will be considered by discussing the effect of interfaces and segregation. Finally one of the essential steps in phase transformations, nucleation, will be dealt with as a means of generating fine particles. Chapter 2 describes the details of nanocrystalline structures and chapters 3–9 describe the

3

manufacture and structure of a number of different types of nanocrystalline metallic materials.

Unstable and metastable equilibria

Nanocrystalline substances are prepared by means of various techniques such as vapour condensation, ball milling and electrodeposition. The particles constituting these materials are usually of sizes as small as a few nanometres; therefore it is legitimate to question whether they possess special thermodynamic properties. However, it can be clearly stated that nanocrystals of the elements display bulk properties in many respects. It was shown in the early 1980s that clusters of some 100–150 atoms exhibit photoelectron spectra characteristic of the bulk metals [4] so their internal energies should have corresponding values (figure 1.1). However, materials prepared by means of the techniques mentioned above are unstable in the as-prepared state mostly because their interfaces are not equilibrated and need to relax to structural states of lower energy. As an example consider high purity elements processed by heavy deformation. Careful thermal analysis shows that recovery may occur even at room temperature, whereas recrystallization takes place at higher temperature. Therefore the material is unstable at room temperature with respect to defect recovery. During recovery, the material samples a series of structural states of

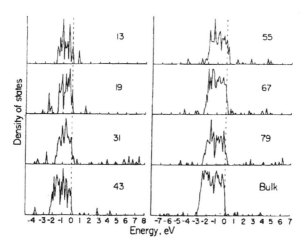

Figure 1.1. Calculated densities of states for clusters of Pd atoms. The digit in each figure 1.refers to the number of atoms in the cluster. The dashed vertical line marks the position of the Fermi level. The bandwidths at half maximum agree with those obtained from x-ray photoelectron spectra of supported metal clusters. The cluster made of 79 atoms has a *d* electron bandwidth 86% of that of bulk Pd. From Baetzold *et al* [3].

progressively lower free energy. It then remains in a metastable state until the temperature is raised to allow recrystallization, the rate of which is enhanced by the presence of non-equilibrium grain boundaries [5].

In order to evaluate the free energy trend, consider the case of a pure element: nanocrystalline Cu obtained by deformation or deposition. The free energy will be taken simply as $\Delta G = \Delta H - T\Delta S$, neglecting second order contributions due to specific heat differences. As for the enthalpy difference, ΔH, high purity Cu, heavily deformed by rolling, gives a single DSC peak of $55\,J/mol$ due to recrystallization at $500\,K$ when heated at $30\,K/min$. If transition elements (e.g. Mn) are added in amounts of parts per million, the recrystallization peak is shifted to higher temperatures (above $673\,K$) and a new peak appears, due to recovery, at temperatures below $473\,K$ [6]. Extrapolating this behaviour to pure Cu, it can be envisaged that recovery of defects should occur both dynamically during deformation and during the storage of samples. Therefore, deformed Cu is unstable at room temperature with respect to defect concentration. Once recovery has occurred, it will remain in a metastable state up to the temperature when recrystallization starts.

Nanocrystalline Cu, prepared as a powder by vapour deposition and compacted, behaves similarly [7]. It releases $300\,J/mol$ around $430\,K$ when analysed immediately after compaction and $53\,J/mol$ around $450\,K$ when analysed five days after preparation. Such values of enthalpy release have been confirmed by a study on nanocrystalline Cu prepared by electrodeposition and cold rolled to variable amounts [8]. On the other hand, nanocrystalline Cu films obtained by sputtering release $1200\text{--}1700\,J/mol$ starting at $450\,K$ [9] and powders prepared by ball milling release $5\,kJ/mol$ [10]. The enthalpy release occurs over a broad temperature range. From these values it was deduced that the interfacial enthalpy may exceed $1\,J/m^2$. Comparison of all data shows that these materials are far from equilibrium soon after preparation, not only because they contain a large amount of interfaces but also because these interfaces are not equilibrated. Their thermodynamic state is ill defined and a series of metastable states may be attained after suitable annealing.

Evaluating the entropy term, ΔS, is not straightforward since there is very little in the literature on the entropy contribution from grain boundaries and interfaces for crystals of any size. Although it is expected that this entropy contribution is small, it can also be conceived that non-equilibrated grain boundaries have a higher entropy. A value of $0.36\,mJ/m^2\,K$ has been estimated for as-prepared nanocrystalline Pt, in contrast to the value for conventional grain boundaries of $0.18\,mJ/m^2\,K$ [11]. Actually the excess entropy per atom sitting in a grain boundary is a substantial part of the entropy of fusion, but the overall entropy per mole of substance sums up to a limited amount even for materials with very small grains.

Using this knowledge of enthalpy and entropy, the free energy of nanocrystalline Cu can be computed taking the bulk single crystal as the reference

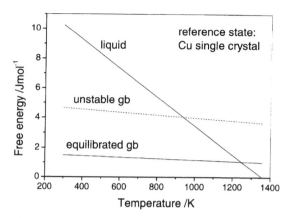

Figure 1.2. The free energy of undercooled liquid and nanocrystalline Cu at different relaxation stages with a reference state of bulk Cu.

state, as shown in figure 1.2. Also shown in figure 1.2 is the extrapolated free energy curve of the undercooled liquid computed from assessed data [12]. The free energy of ball milled Cu apparently accounts for a substantial fraction of the free energy of melting whereas the free energy of the annealed state does not exceed 20% of the free energy of melting. In between there is a whole series of accessible states, unstable with respect to the defect concentration, spanning a wide range of free energies.

Curvature effects

Once the interfaces are equilibrated, the excess contribution to the free energy ΔG of a nanostructered material with respect to the same substance having large grains can be described by means of the Gibbs–Thomson equation

$$\Delta G = \frac{2\gamma V_{\mathrm{m}}}{r} \tag{1.1}$$

where γ is the interfacial free energy, V_{m} the molar volume and r the radius of the grains.

The excess free energy has an effect on various phenomena. If it is considered relative to the free energy of fusion ($\Delta G \cong \Delta S_{\mathrm{m}}(T_{\mathrm{m}} - T_{\mathrm{m}}(r))$) it will cause lowering of the melting point, T_{m}. At a given grain radius r, the melting point $T_{\mathrm{m}}(r)$ is

$$T_{\mathrm{m}}(r) = T_{\mathrm{m}} - \frac{2\gamma_{\mathrm{sl}}}{\Delta S_{\mathrm{m}} r} \tag{1.2}$$

where γ_{sl} is the interfacial free energy difference between liquid and crystal and ΔS_{m} is the entropy of fusion. This effect has been demonstrated with

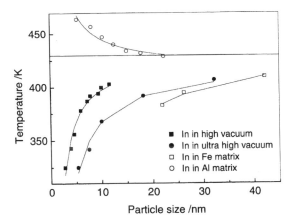

Figure 1.3. The dependence of melting point on size for In nanocrystals in various matrixes. Full symbols refer to supported pure In particles in different vacuum conditions. Open symbols refer to In particles embedded in a metal matrix. The lines through the points are fits according to equation (2). Adapted from Shi [15]. The horizontal line marks the equilibrium melting point of bulk In.

small particles embedded in inert matrices [13]. However, it should be noted that further energy contributions may easily overcome the one expressed by equation (1.2), therefore reversing the effect, as shown in figure 1.3. Particles contained in a crystalline matrix can solidify coherently with the matrix, i.e. with the same crystallography. As a consequence the interfacial free energy difference between liquid and crystal is strongly modified and may become formally negative as testified by an increase of the melting point instead of a decrease. Examples, e.g. Pb in Al, are collected in [14].

There have been several attempts to explain findings on lowering and enhancing the melting point of particles constrained in a matrix. Equation (1.2) expressing the capillarity effect is modified to

$$T_m(r) = T_m - \frac{3\left(\dfrac{\gamma_{sm}}{\rho_s} - \dfrac{\gamma_{sl}}{\rho_l}\right)}{\Delta S_m r} \tag{1.3}$$

where γ_{sm} is the interfacial free energy difference between matrix and crystal and ρ_s and ρ_l are the densities of the solid and liquid respectively. The embedded particles can experience a pressure variation due to differential thermal expansion of the matrix and the particles themselves and due to volume changes on melting. The pressure variation contributes to the change in melting point according to the Clausius–Clapeyron equation. There can be, however, relief of the pressure by formation of vacancies and creep. An alternative view considers that the vibrational entropy of surface atoms is increased with respect to those of the bulk [15]. The model implies an increase in pressure on the surface atoms during melting. However, to obtain the pressure at the interface one would need to know the interfacial energy and its dependence on curvature. Therefore, although some insight has been gained, several fundamental questions remain unanswered, the central one being the eventual change in interfacial energy in very fine particles.

In a multicomponent material containing fine particles, ΔG from equation (1.1) will affect the chemistry of the system of matrix and embedded crystals. The solute concentration in the vicinity of the small particles must be enhanced with respect to that in the bulk of the matrix. If for simplicity the solution is assumed to behave ideally, then $\Delta G = -RT \ln(x_{i(r)}/x_{i(r=\infty)})$ and equation (1.3) becomes

$$ \ln\left(\frac{x_{i(r)}}{x_{i(r=\infty)}}\right) = \frac{V_i}{RT}\frac{2\gamma_{\mathrm{sl}}}{r} \qquad (1.4) $$

where V_i is the molar volume and $x_{i(r)}$ are solute molar fractions. This manifestation of the Gibbs–Thomson effect has been used to estimate γ_{sl} in studies on the formation of nanocrystalline Al crystals in Al-RE-TM (RE = rare earth and TM = transition metal element) glasses where primary precipitation of fcc Al occurs [16]. The volume fraction of precipitated phase determines the concentration, $x_{\mathrm{Al}(r)}$, remaining in the matrix surrounding the fine crystals and can be determined via macroscopic measurements of the heat release during transformation or the total intensity of x-ray reflections in diffraction patterns.

Stabilization of grain boundaries

There has been much controversial discussion on the possibility that some grain boundary configurations correspond to a relative minimum in free energy in nanocrystalline materials. It was proposed that the excess free volume in the grain boundaries results in a high interfacial entropy stabilizing the material [17]. On the other hand, it has been suggested that a stabilization of grain boundaries in milled elements could occur at very low grain sizes due to a decrease in grain boundary free volume, the occurrence of which was inferred from the scaling of resistivity with grain size [18]. This effect should be caused by the interaction between grain boundaries mediated through strain fields. Attempts to determine the excess entropy by means of measurements of the electromotive force of nanocrystalline electrodes with respect to coarse-grained materials at various temperatures were equally controversial in that they provided scattered results according to the relaxation state of the grain boundaries [19]. It was finally recognized that the structure of nanocrystalline grain boundaries depends on the preparation technique as shown in figure 1.4 and, therefore, grain boundaries

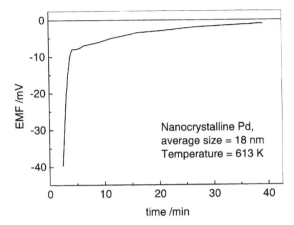

Figure 1.4. The electromotive force of an electrode made of nanocrystalline Pd (grain size 18 nm) tested against an electrode of large-grained Pd. The electromotive force is always negative, corresponding to a positive Gibbs free energy difference. The time dependence of the electromotive force is related to structural changes. The grain size increases significantly after measurement at the given temperature due to the instability of the as-prepared nanocrystalline material.

formed under highly non-equilibrium conditions are generally unstable and structurally different [20].

Stabilization can arise from the presence of atomic segregation at grain boundaries which is described by the Gibbs adsorption isotherm

$$\frac{\partial \gamma}{\partial \ln x_s} = -RT\,\Gamma_s \tag{1.5}$$

where x_s is the solute molar fraction, R is the gas constant, T is the absolute temperature and Γ_s is the number of solute atoms per unit area of boundaries or the surface excess. From equation (1.5), if the surface excess is positive, γ must decrease with increasing solute concentration and there is less driving force for grain growth. For a given solute content, however, the number of solute atoms per unit area of grain boundaries Γ_s decreases with decreasing particle size. Consequently, γ should be progressively less affected. These various possibilities may have occurred in different cases, such as deposition of clusters from the vapour, ball milling or severe plastic deformation of bulk metals, so that a spread in values of γ has been found. The segregation of solutes at grain boundaries represents a major effect in multicomponent systems.

The idea has been put forward to include the interaction of chemical components with topological defects [21]. The resulting prediction is that there can exist a state for which the polycrystal is stable with respect to variation of its grain boundary area and therefore grain size reaching a local minimum in free energy. It should be noted that this extension does not include curvature effects.

Nanostructures and amorphization

Reaching a fine grain size in alloys could mean loss of periodic order and therefore amorphization. Although experimentally the border between extremely fine microstructures and formation of amorphous solids may be sometimes difficult to set, in an idealized experiment a crystal can be sub-divided to such an extent that the melting point is lowered and would reach the glass transition where the crystal would become inherently unstable and would transform to a glass [3]. An alternative formulation of this concept is that the crystal would reach a size lower than that of a critical nucleus and would spontaneously melt, as justified in the next paragraph.

Thermodynamics of nucleation

A particle embedded in a matrix must have originated from an assembly of atoms called a nucleus, which grows by progressively consuming the matrix. Since the nucleus is small, it may be alternatively called a nanocrystal. Nucleation is required for first-order transformations, i.e. those transformations implying a discontinuity in the temperature or pressure derivatives of the free energy of the system. The new phase has a structure differing from that of the matrix, therefore a structural discontinuity must occur at the site where it forms. The classical theory developed for nuclei formed in fluids or glasses accounts for the free energy difference between nucleus and matrix ΔG_c and the interfacial energy γ. The nucleus has a critical radius, r_c:

$$r_c = -\frac{2\gamma}{\Delta G_c}. \tag{1.6}$$

A cluster smaller than r_c will tend to disappear (figure 1.5). Note that equations (1.1) and (1.5) are essentially the same, i.e. the critical radius of the nucleus equals that of a small particle at its melting point or at the limit of dissolution in the matrix. ΔG_c and γ are the key parameters which define also the driving force for the process. The formation of a new phase in the solid state would involve at least a further contribution to the system free energy, such as a term expressing the mechanical energy due to the stress on either the matrix or the nucleus arising from the change in specific volume during transformation.

    The case of a liquid or glassy matrix is considered here. The description of ΔG_c in polymorphic and eutectic transformations, when on average there

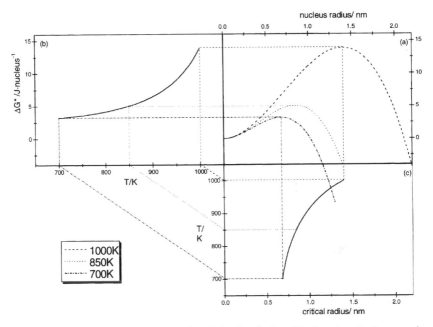

Figure 1.5. The activation barrier ΔG^* and the size r^* of a critical nucleus in the case of a model system at various temperatures. In (a) the free energy of formation of the nucleus is shown as a function of size. It goes through a maximum corresponding to the nucleus of critical size. The free energy of the system decreases both when particles finer than the critical ones dissolve and when particles larger the critical size grow. On increasing the undercooling, both ΔG^* and r_c decrease steadily, therefore the size can be tuned by controlling the nucleation process and limiting the subsequent growth.

is no change in composition between the new crystals and the matrix, has seen much progress. It is computed using enthalpy and entropy contribution according to

$$\Delta H = \Delta H_m - \int_{T}^{T_m} T \, \Delta C_p \, dT \quad \text{and} \quad \Delta S = \Delta S_m - \int_{T}^{T_m} \Delta C_p \, \frac{dT}{T} \quad (1.7)$$

where ΔH_m is the melting enthalpy and $\Delta C_p = C_p^l - C_p^s$ is the difference between the specific heat of the two phases.

As dictated by equation (1.7), the temperature behaviour of ΔG depends on the temperature behaviour of ΔC_p [22]. The specific heat is known for metals and some alloys above the melting point, but there are only rare data in the undercooled regime. For glass-forming systems there have mostly been experiments at temperatures in the neighbourhood of the glass transition T_g. Approximations for ΔC_p have been recently reviewed showing that ΔG can be reproduced with good confidence [23, 24].

Primary transformations implying compositional partition are more difficult to model [25]. They are, however, the most interesting for metallic

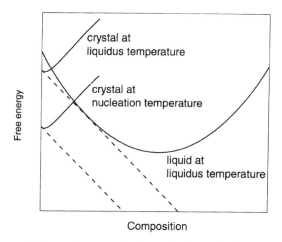

Composition

Figure 1.6. The parallel tangent construction for nucleation of a phase within a matrix (here taken as a highly undercooled liquid or a glass). At the liquidus equilibrium temperature the common tangent to the free energy curves defines the composition of the coexisting phases. If the liquid is undercooled the most probable nucleus forms at a composition defined by the tangent to the solid free energy curve which is parallel to the previous tangent.

materials [26]. In fact, bulk nanostructured alloys are obtained by controlled heat treatment of either glasses or supersaturated solid solutions [27]. The stability of such microstructures with respect to coarsening is very much related to solute concentration gradients established in the matrix after precipitation. As mentioned above, the solute build-up at nanocrystal interfaces may be used to get an estimate of the interfacial energy according to equation (1.4) [16].

ΔG_c stems from the parallel tangent construction shown in figure 1.6, where the composition of the most probable nucleus is found by drawing the tangent to the liquid free energy curve at a given alloy composition and then finding the parallel line tangent to the solid free energy curve. ΔG_c can also be obtained from empirical formulae. The simplest formula is

$$\Delta G = \Delta \mu_A = (T_l - T)(\Delta S_A^m - R \ln x_A^l) \qquad (1.8)$$

where $\Delta \mu_A$ is the difference in chemical potential for the A component between the nucleus and the matrix, T_l is the liquidus temperature and x_A^l is the A concentration in the matrix. The technique of computing phase diagrams can be extended to include metastable equilibria, implying the calculation of free energy functions for phases in metastable state, e.g. in the undercooled regime. This provides the means for obtaining the driving forces for nucleation. The most general expression for $\Delta \mu_A$ is

$$\Delta \mu_A = d\, G_{A,f}(T) + RT \ln \frac{x_{A,l}}{x_{A,s}} + \mu_{A,l}^e(x_{A,l}, T) - \mu_{A,s}^e(x_{A,s}, T) \qquad (1.9)$$

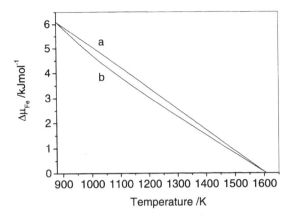

Figure 1.7. The driving force for primary crystallization of α-Fe crystals from an Fe-Zr liquid or amorphous alloy as computed (a) with equation (8) and (b) by means of the CALPHAD procedure. The driving force is expressed as difference in chemical potential of the precipitating element. It is nil at the liquidus temperature and becomes substantial with increasing undercooling. The upward bending of the free energy obtained in (b), rather unusual for such plots, is due to the magnetic contribution which is not accounted for in (a).

where $\Delta G_{A,f}$ is the free energy of melting of component A at temperature T, $x_{A,s}$ is the nucleus composition and $\mu_{A,I}^{e}$ is the excess chemical potential for the A component in the I phase. The quantities $\Delta G_{A,f}$ and $\mu_{A,I}^{e}$ are obtained as outputs in calculations of phase diagrams (CALPHAD) (see figure 1.7). In the calculations, the thermodynamic quantities and the phase equilibria are assessed with multiparametric fitting procedures by using available experimental data. It is a relatively new implementation of these quantitative thermodynamic methods to provide input data on transformation of metastable phases (e.g. temperatures and enthalpies of crystallization of a glass) in order to compute more reliable free energies of undercooled phases.

The interfacial energy between crystal and melt is known experimentally for a number of elements at their melting point, but often with large uncertainties. Very few data are available for binary compositions and compounds. In the undercooled regime γ is not an equilibrium quantity. It is an operative parameter expressing the work needed for crystal nucleation.

The simplest models are based on evaluation of the difference in the interaction energy of nearest next-neighbours between the crystal and the melt which accounts for the interface excess energy. A recent development of this model gives for the interfacial energy [28]

$$\gamma_{sl} = \left[\left(\frac{V_{l,m}}{V_{s,m}} \right)^{2/3} - 1 \right] \gamma_{lv} + \frac{0.25 \cdot \Delta H_m + 0.19 \cdot T \cdot \Delta S_m}{N_{Av}^{1/3} V_{s,m}} \qquad (1.10)$$

where γ_{lv} is the liquid/vapour interfacial energy, $V_{s,m}$ and $V_{l,m}$ are the solid and liquid molar volumes respectively and N_{Av} is the Avogadro constant. Most values for metals computed at the melting point ($T = T_m$) agree well with experimental results. On undercooling, γ_{sl} decreases as suggested by the limited number of experimental data and models for the work of formation of a critical nucleus from the melt which consider a finite thickness of the interface instead of taking it as a purely geometrical surface [29, 30]. These diffuse interface models give an estimate of the temperature dependence of the ratio $\gamma_{sl}(T)/\gamma_{sl}(T_m)$, which decreases smoothly with decreasing temperature. A linear fit to the curves reported in [29] and [30], gives

$$\gamma_{sl}(T)/\gamma_{sl}(T_m) = 0.48 + 0.52(T/T_m) \tag{1.11}$$

which agrees reasonably with equation (1.10).

The expressions derived in various models remain valid as long as the assumption of a flat interface represents the status of a real interface. They may not be applicable to extremely fine particles. Therefore further modelling of interfacial energies is needed for improved description of crystal formation via nucleation.

Soft chemistry nanostructures

Molecular organization can be achieved either via spontaneous self-assembly of components or by templating definite structures with suitable substrates [1]. The single supramolecular units may be considered equilibrium substances as most organic and metallorganic compounds. On the other hand they are building blocks of solids containing cages, cavities, channels etc. and are, therefore, likely to be metastable with respect to close packing. These solids can also present reversible phase transformations such as those occurring in liquid crystals.

Inorganic mesoporous materials obtained by templating, e.g. silicalite, contain cages and channels which are regularly distributed, i.e. periodic, and formed with walls of amorphous silica. Here again it can be envisaged that the structural state represents a local minimum in a free energy landscape. In fact, they are best described as metastable in that the structures could collapse to bulk silica phases via suitable thermal treatment. Nevertheless, the periodic arrangement of such structures can display reversible phase transitions caused by slight rotation of the silica building blocks [31].

References

[1] Gleiter H 2000 *Acta Mater.* **48** 1
[2] Battezzati L 1997 *Mater. Sci. Forum* **235–238** 317

[3] Loikowski W and Fecht H-J 2000 *Prog. Mater. Sci.* **45** 339

[4] Baetzold R C, Mason M G and Hamilton J F 1980 *J. Chem. Phys.* **72** 366;
     Lee S-T, Apai G, Mason M G, Benbow R and Hurych Z 1981 *Phys. Rev. B* **23** 505

[5] Lian J, Valiev R Z and Baudelet B 1995 *Acta Mater.* **43** 4165

[6] Lucci A, Riontino G, Tabasso M C, Tamanini M and Venturello G 1978 *Acta Metall.* **26** 615

[7] Günther B, Kupmann A and Kunze H-D 1992 *Scripta Metall. Mater.* **27** 833

[8] Lu L, Sui M L and Lu K 2001 *Acta Mater.* **49** 4127

[9] Huang Y K, Menovsky A A and de Boer F R 1993 *NanoStruct. Mater.* **2** 587

[10] Eckert J, Holzer J C, Krill C E III and Johnson W L 1992 *J. Mater. Res.* **7** 1751

[11] Tschöpe A and Birringer R 1993 *Acta Metall.* **41** 2791

[12] Dinsdale A T 1991 *CALPHAD* **15** 317

[13] Kim W T and Cantor B 1992 *Acta Metall. Mater.* **40** 3339

[14] Chattopadhyay K and Goswami R 1997 *Prog. Mater. Sci.* **42** 287

[15] Shi F G 1994 *J. Mater. Res.* **9** 1307;
     Jiang Q, Zhang Z and Li J C 2000 *Acta Mater.* **48** 4791

[16] Jiang X J, Zhong Z C and Greer A L 1997 *Mater. Sci. Eng. A* **226–228** 789

[17] Fecht H-J 1994 in *Fundamental Properties of Nanostructured Materials* ed. D Fiorani
     and G Sberveglieri (Singapore: World Scientific) pp 151–171

[18] Krill C E and Birringer R 1996 *Mater. Sci. Forum* **225** 262

[19] Kircheim R, Huang X Y, Cui P, Birringer R and Gleiter H 1992 *NanoStruct. Mater.*
     1 167

[20] Gärtner F, Bormann R, Birringer R and Tschöpe A 1996 *Scripta Mater.* **35** 805

[21] Weissmüller J 1993 *NanoStruct. Mater.* 3 261

[22] Battezzati L and Garrone E 1984 *Z. Metallkde.* 75 305

[23] Kelton K 1991 in *Solid State Physics* vol 45, ed. H Ehrenreich and D Turnbull (New
     York: Academic Press) p 75

[24] Battezzati L and Castellero A 2002 *Nucleation and the Properties of Undercooled
     Melts, Materials Science Foundations* vol 15 ed. M Magini and F H Wohlbier
     (Uetikon-Zurich, Switzerland: TransTech Publications) pp 1–80

[25] Battezzati L 1994 in *Fundamental Properties of Nanostructured Materials* ed.
     D Fiorani and G Sberveglieri (Singapore: World Scientific) pp 172–189

[26] Battezzati L 1996 *Mater. Res. Soc. Symp Proc.* **400** 191

[27] Inoue A and Kimura H 2001 *J. Light Metals* **1** 31

[28] Battezzati L 2001 *Mater. Sci. Eng. A* **304–306** 103

[29] Spaepen F 1994 in *Solid State Physics* vol 47, ed. H Ehrenreich and D Turnbull (New
     York: Academic Press) p 1

[30] L Granasy 1993 *J. Non-Crystalline Solids* **162** 301–303

[31] Bell R G, Jackson R A and Catlow C R A 1990 *J. Chem. Soc. Chem. Comm.* **10** 782

Chapter 2

Nanostructure of amorphous alloys

Yoshihiko Hirotsu, Tadakatsu Ohkubo
and Mitsuhide Matsushita

Introduction

High resolution electron microscopy (HREM) has been used to understand the local structure of amorphous alloys [1–6], which are strongly dependent on their quenching history in forming the amorphous structure [5]. Atomic medium range order (MRO) [7] in which atomic correlation extends more than several atomic distances often exists in amorphous alloys formed by rapid-quenching and sputter-deposition. The structures of these medium range order regions can be estimated by high-resolution electron microscopy and image simulation when the medium range order sizes are as large as 1–2 nm on the basis of local lattice-fringe spacings and cross-angles of fringes [4–6, 8]. In amorphous $Fe_{84}B_{16}$ body centred cubic medium range ordered structures were identified, while in amorphous $Pd_{77.5}Cu_6Si_{16.5}$ [5] and $Pd_{82}Si_{18}$ [6] face centred cubic medium range ordered structures were identified in the as-formed states by high-resolution electron microscopy. The high-resolution electron microscopy identification of the structure, however, becomes uncertain when the atomic arrangements of medium range order become complex. In such cases, a nanoelectron probe with a probe size of 0.5–2 nm in a field-emission transmission electron microscope (FE-TEM) helps to produce electron diffraction structure analysis of the medium range order [8].

In addition to medium range order, atomic short range order (SRO) which defines the nearest-neighbour atomic arrangement [7] is important to understand atomic correlations (averaged) and the correlation distances in amorphous alloys. This can be achieved by atomic pair distribution function (PDF) analysis using x-ray, neutron and electron diffraction techniques [7]. Electron diffraction has an advantage in studying atomic correlations concerned with lighter atoms and in obtaining information up to high scattering angles. Local pair distribution analysis from nanometre areas combined with transmission electron microscope observation is another advantage. Recent advances in electron intensity recording systems like imaging-plate (IP) and slow-scan CCD have made it possible to measure electron intensities precisely for the purpose of electron diffraction intensity analysis. Recent electron-energy-loss spectroscopy (EELS) and energy-filtering techniques which can remove the inelastic part of intensity also helps the diffraction intensity analysis. These new techniques in addition to nanodiffraction and high-resolution electron microscopy are becoming necessary for developing comprehensive understanding of the atomic structures of non-equilibrium materials.

This chapter describes high-resolution electron microscopy studies, using combined nanoprobe, image plate, electron energy loss spectroscopy and energy-filtering techniques, to determine atomic pair distribution functions and local atomic short and medium range order. The chapter reviews such structural analyses mainly for amorphous Pd-Si [8-11], Fe-Zr-B [12] La-Al-Ni [13] and Zr-Pd [14]alloys. Nanoscale phase separations are expected to occur in these amorphous alloys and to explain as-formed and annealed structures. Chapters 3–9 give more details of the manufacture and structure of a number of different nanocrystalline metallic materials.

Electron microscopy

Thin amorphous $Pd_{82}Si_{18}$ and $Pd_{75}Si_{25}$ films with a thickness of about 10 nm are deposited on NaCl substrates in an Ar-beam sputtering equipment with a cryo-pump evacuation system, and a substrate cooled below room temperature during sputtering. For amorphous $Fe_{90}Zr_7B_3$, $La_{55}Al_{25}Ni_{20}$ and $Zr_{70}Pd_{30}$ alloy specimens, amorphous ribbons with a thickness of about 20 nm are formed by rapid-quenching, followed by ion-thinning with a cold stage for transmission electron microscopy.

Nanoprobe electron diffraction has been performed using 200 kV JEM-2010F and 300 kV JEM-3000F field emission gun transmission electron microscopes. Diffraction patterns are recorded using an image plate and CCD-camera video system. Selected-area electron diffraction (SAED) patterns for pair distribution function analysis has been performed using 200 kV JEM-2010F and JEM-2010 transmission electron microscopes. To decrease multiple scattering, thin specimen areas are selected for the Fe-Zr-B, La-Al-Ni and Zr-Pd alloys. The inelastic part of the intensity is removed using energy-filters (in-column or post-column type) or estimated from electron energy loss spectroscopy. Diffraction intensity profiles are recorded on an image plate. In both nanoprobe and selected area diffraction, camera-lengths are corrected using reference diffraction patterns from fine gold particles. Structural observations have also been performed by high-resolution electron microscopy and specimen heating experiments by conventional transmission electron microscopy. For medium range order observation by high-resolution electron microscopy, the defocus method [8] takes the high contrast imaging conditions into consideration with respect to the phase-contrast transfer function [15].

Medium range order

As already found in vapour-quenched [6] and rapid-quenched [8] specimens of Pd-Si amorphous alloys with near-eutectic compositions, medium range ordered regions with sizes of about 1–2 nm are frequently observed as localized lattice fringe images. From the fringe spacings and crossed-fringe geometries, the structure is judged to be fcc Pd. As an example, the medium range ordered regions in an amorphous $Pd_{82}Si_{18}$ film are clearly seen in figure 2.1 in the encircled areas. With a probe size of about 1–2 nm, nanodiffraction patterns were taken from the amorphous $Pd_{82}Si_{18}$ film [11]. The diffraction patterns can only be explained by an fcc Pd structure. In the insets in figure 2.1, fcc patterns from the medium range order regions are seen. Also in the insets, a halo-diffraction pattern by selected area diffraction is shown for comparison. The structure of the medium range order in amorphous Pd-Si near the eutectic composition is therefore the same as the α phase that forms during primary crystallization.

When the Si composition of the amorphous Pd-Si alloy increases to the compound composition of Pd_3Si, the characteristic features of the medium range order images become different from fcc. Figure 2.2 shows a high-resolution electron microscope image [10] of an amorphous $Pd_{75}Si_{25}$ film, where crossed-lattice fringe regions of medium range order are seen in the

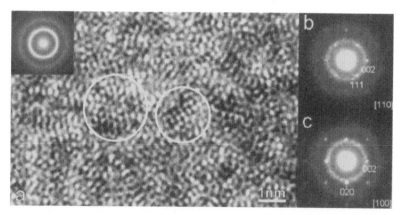

Figure 2.1. (a) HREM image of α-$Pd_{82}Si_{18}$ thin film with local MRO regions (encircled). SAED pattern is in the inset of (a). (b) and (c) are nanodiffraction zone-axis patterns from the fcc Pd(Si) MRO regions with indices of [110] and [100], respectively.

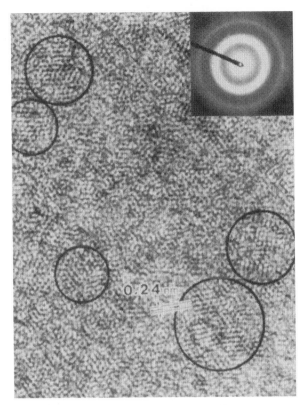

Figure 2.2. HREM image of an amorphous $Pd_{75}Si_{25}$ alloy thin film. MRO regions with clear crossed-lattice fringes are encircled. In spite of the dense formation of MRO regions, a halo-diffraction pattern is taken by SAED (inset).

encircled areas. In this case, nanodiffraction analysis becomes effective, and nanodiffraction patterns from the medium range order regions are not of fcc Pd structure but of a lower symmetry structure. Examples of nanodiffraction single crystal patterns from the medium range order regions are shown in figure 2.3. By analysing various zone-axis patterns, the reciprocal lattice and nanodiffraction intensities of the medium range order can be explained as similar to the hexagonal Pd_2Si-type structure (space group: $P\,6\,2m$) [16]. The lattice parameters are $a = 0.715$ and $c = 0.312\,nm$. The images cannot be explained by the well-known orthorhombic Pd_3Si structure. Hexagonal Pd_2Si is known to be synthesized stably at lower heating temperatures compared with Pd_3Si [17], and has been confirmed to appear in the primary crystallization stage of the amorphous $Pd_{75}Si_{25}$ alloy [11].

Figure 2.4 shows the atomic arrangements of the Pd_2Si-type medium range order. The atomic positional parameters and Si occupancies are determined so that calculated intensities of diffraction spots agree with the

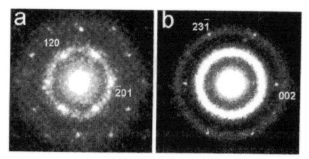

Figure 2.3. Zone-axis nanodiffraction patterns from MROs in the $Pd_{75}Si_{25}$ alloy thin film. Reflection indices are based on the Pd_2Si-type structure.

observed relative diffraction intensities (classified into five levels) [10]. Two types of trigonal Pd prism columns are running along the *c*-axis. The average Pd-Pd bond distance of the medium range order structure is 0.30 nm and is slightly larger than that of the average Pd-Pd distance, 0.28 nm, measured from the first main peak of the pair distribution function of this alloy. This means that an atomic density fluctuation must be occurring in amorphous $Pd_{75}Si_{25}$ with respect to the formation of the medium range order. Almost the same result is obtained for the medium range order of an amorphous $Pd_{76}Si_{24}$ ribbon [18].

It was difficult to record nanodiffraction single crystal patterns from the medium range ordered regions in other amorphous alloys observed by high-resolution electron microscopy, because of the medium range ordered regions are too small, <1 nm, compared with amorphous Pd-Si alloys.

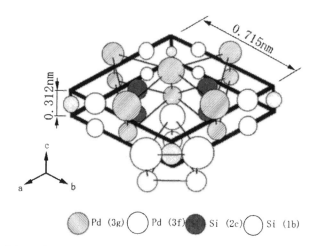

Figure 2.4. Schematic representation of the hexagonal Pd_2Si-type structure.

Nano-scale phase separation

$Pd_{82}Si_{18}$ and $Pd_{75}Si_{25}$

In the previous section, medium range order structures in amorphous $Pd_{82}Si_{18}$ and $Pd_{75}Si_{25}$ alloys reflect the structures of their primary crystallization in the decomposition processes. The amorphous $Pd_{82}Si_{18}$ alloy structure has been investigated further from the viewpoint of phase separation of the alloy on a nanometre scale with respect to medium range order formation [11].

In order to prove the occurrence of nanoscale phase separation in amorphous $Pd_{82}Si_{18}$, a structure model is constructed where fcc Pd atomic clusters including Si are embedded in a dense-random-packed (DRP) structure of Pd and Si, and the calculated pair distribution function based on the model structure is compared with that obtained from a selected area diffraction experiment. For structure modelling, four initial basic structures are constructed, with different medium range order fractions $A_f = 0$ (dense random packed), 0.1, 0.2 and 0.3. A_f is the atomic fraction of total medium range order atoms in the structural volume considered. Two medium range order Si contents of 9 and 18 at% are considered. The shape of the medium range order is chosen to be spherical with a diameter of 1.9 nm (averaged size evaluated from high-resolution electron microscopy). The medium range ordered regions are randomly distributed with random orientations in a structural cell ($3.305 \times 3.305 \times 3.305$ nm) under the defined A_f. The intervening open space between the medium range ordered regions is then filled with Pd and Si randomly by taking both the total atomic concentration and the density into consideration. The resulting medium range order and matrix structures are then relaxed energetically using Lennard-Jones (L-J) potentials. The density of the model structure is chosen as 10.61 g/cm^3 by interpolating the experimental densities of the amorphous Pd-Si alloys [19]. The total number of atoms for all the initial structures is 2500.

Pair distribution function

According to diffraction theory [7], a reduced interference function $F(Q)$ is defined as

$$F(Q) = [I(Q) - N\langle f^2 \rangle]QN\langle f \rangle^2 = [I_{obs}(Q) - BG(Q)]Q\langle f^2 \rangle BG(Q)\langle f \rangle^2 \tag{2.1}$$

where $I(Q)$ and $I_{obs}(Q)$ are the actual and observed elastic scattering intensities, r is the atomic radial distance, N is the total number of atoms, $Q = (4\pi/\lambda)\sin\theta$ is the scattering vector, θ is the half scattering angle, λ is the electron wave length, and $BG(Q)$ the background intensity which smoothly links the middle points between the intensity maxima and minima of the halo-intensity profile almost along the $\langle f \rangle^2$ curve. The square-mean and mean-square atomic scattering factors for electrons, $\langle f^2 \rangle$ and $\langle f \rangle^2$, are expressed as $\langle f^2 \rangle = N_j f_j / N$ and $\langle f \rangle^2 = (N_j f_j)^2 / N^2$, where N_j and f_j are the atomic number and atomic scattering factor (including the temperature factor) for element j, respectively. The reduced distribution function $G(r)$, radial distribution function $RDF(r)$, and atomic pair distribution function $g(r)$ can be obtained from $F(Q)$ by Fourier transformation. It is important to note that, for precise electron diffraction pair distribution function analysis, the most important procedure is to obtain diffraction intensity profiles under conditions as close as possible to kinematical diffraction, with insignificant inelastic scattering or multiple scattering, and with good linearity in the intensity recording. For kinematical conditions, the background intensity $BG(Q)$ lies almost along the $\langle f^2 \rangle$ curve.

The following procedure is used to obtain the experimental pair distribution function. Selected area diffraction patterns are taken from amorphous $Pd_{82}Si_{18}$, and recorded on an image plate. The intensity profile is averaged

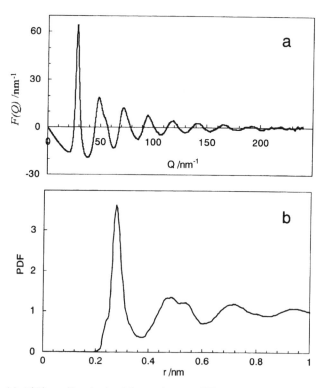

Figure 2.5. (a) $F(Q)$ profile obtained from electron diffraction intensity analysis for α-$Pd_{82}Si_{18}$ thin film. (b) PDF profile obtained from the function in (a).

along the radial direction in reciprocal space up to a scattering vector $Q = 240\,\mathrm{nm}^{-1}$. The inelastic intensity is estimated from electron energy loss spectroscopy spectra by measuring the fraction of energy-loss electrons as a function of scattering angle [9]. The reduced interference function $F(Q)$ can then be obtained from equation (2.1) and Fourier transformed to give the reduced radial distribution function and pair distribution function $g(r)$. Figures 2.5(a) and (b) show the reduced interference function $F(Q)$ and pair distribution function $g(r)$ for an amorphous $Pd_{82}Si_{18}$ thin film. At $r = 0.242\,\mathrm{nm}$ in $g(r)$ there appears a small subsidiary peak related to the Pd-Si correlation. The first main peak corresponds to the Pd-Pd correlation peak with a Pd-Pd distance of 0.275 nm. The resulting coordination number for Pd-Pd is 12.1. The Si-Pd (central Si) coordination number is not easy to obtain from the subsidiary Pd-Si peak, because of the difficulty in separating it from the tail of the main Pd-Pd peak.

Monte-Carlo simulation

Reverse Monte Carlo (RMC) simulation [20] can be used to determine a good fit structural model and explain the experimental pair distribution function profiles. The criterion for convergence of the simulation is [20]

$$\chi^2 = \sum (g_{\mathrm{exp}}(r) - g_{\mathrm{cal}}(r))^2 N \qquad (2.2)$$

where $g_{\mathrm{exp}}(r)$ and $g_{\mathrm{cal}}(r)$ are experimental and reverse Monte Carlo calculated pair distribution functions and N is the number of data points. To obtain a good fit atomic structure by reverse Monte Carlo simulation, χ^2 is minimized using at least 10^5 iterations. Figure 2.6 shows an example of reverse Monte Carlo fitting to the experimental pair distribution function

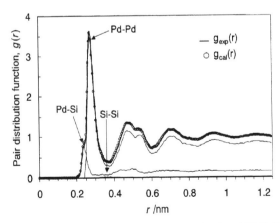

Figure 2.6 Profile fitting of experimental PDF by RMC-simulated PDF. Partial PDFs are also simulated.

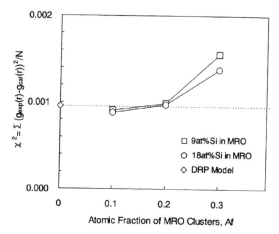

Figure 2.7. Final mean square deviations against MRO atomic fraction A_f for structure models of α-$Pd_{82}Si_{18}$ with different Si contents in MRO.

for a model structure (medium range order fraction $A_f = 0.1$ and medium range order Si content $= 18$ at%). The fitting of $g_{cal}(r)$ to $g_{exp}(r)$ is very good, and the partial pair distribution functions for the Pd-Pd and Pd-Si correlations can be obtained as drawn in the figure. The final χ^2 values for different initial structural models are plotted in figure 2.7 as a function of A_f. No medium range order, $A_f = 0$, in amorphous $Pd_{82}Si_{18}$ corresponds to dense random packing. Four other structures are near dense random packings, with medium range order fractions of $A_f = 0.1$ and 0.2 and Si contents of 9 and 18 at%.

Voronoi polyhedral analysis [21] can be performed on the resulting good fit structural model. Figure 2.8 shows Voronoi polyhedral analysis of the matrix dense random packed structure for a reverse Monte Carlo simulated structural model with medium range order fraction $A_f = 0.1$ and Si content 18 at%. The figure shows the types of polyhedra with their Voronoi indices and fractional contents. The analysis takes two nodes of a Voronoi polyhedron as the same when their separation is less than 0.01 nm. Trigonal-prism-related polyhedra are frequently observed, consistent with the results of x-ray and neutron diffraction studies [22] of amorphous Pd-Si alloys. The medium range order structures and pair distribution function profile of amorphous $Pd_{80}Si_{20}$ melt spun ribbons also show similar results.

$Fe_{90}Zr_7B_3$

From the above results, a structural feature of α phase clusters embedded in a dense random packed matrix must be common in amorphous alloys with near-eutectic compositions. As in eutectic amorphous Fe-B [4], medium range ordered regions as small as 1nm which have been identified as bcc

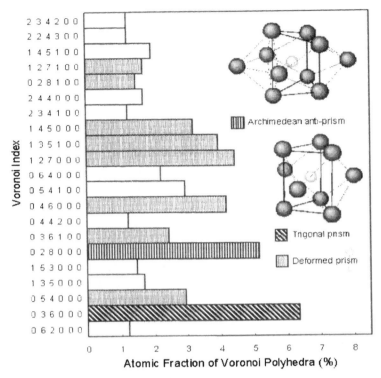

Figure 2.8. Frequency histogram of Voronoi polyhedra in the matrix structure of the simulated structure model for α-$Pd_{82}Si_{18}$. In the model the Si content in MRO and A_f are 18 at% and 0.1, respectively. Atomic arrangements of trigonal and Archimedean antiprism structures are schematically drawn in the figure.

by high-resolution electron microscopy in an amorphous $Fe_{90}Zr_7B_3$ alloy [12], although nanodiffraction study was not possible to identify the structure due to the smallness of the medium range order regions. An example of the high-resolution electron microscopy images of this alloy is shown in figure 2.9, where small medium range order regions with bcc [100] and [111] cluster images are visible in the encircled areas. In order to know the more detailed local structure of this amorphous alloy, a structure modelling study has been performed [12], where a reasonable structure model is made to realize the experimental pair distribution function profile using the reverse Monte Carlo simulation and Voronoi analysis. A selected area diffraction pattern from the amorphous alloy is recorded on an image plate using an energy-filter transmission electron microscopy, followed by pair distribution function analysis in a similar manner to the structure analysis of Pd-Si described above. The reverse Monte Carlo simulation of the pair distribution function started from an initial dense random packed

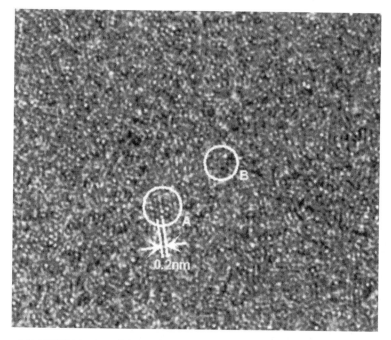

Figure 2.9. HREM image of the as-formed α-$Fe_{90}Zr_7B_3$. [111] and [100] images of bcc clusters are visible in areas A and B, respectively.

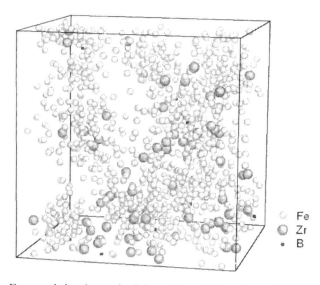

Figure 2.10. Fe atoms belonging to the deformed-bcc clusters observed in the RMC-simulated structure model for $Fe_{90}Zr_7B_3$.

structural model with 2500 atoms, made by relaxing energetically the structure using Lennard-Jones potentials for the constituent atoms. The Lennard-Jones parameters were determined using the melting points of Fe, Zr, Fe_2Zr, FeB and B_2Zr. Local atomic configurations are obtained by Voronoi polyhedral analysis after finishing the reverse Monte Carlo simulation. From the Voronoi analysis, deformed-bcc Fe clusters with indices such as '036400' and '028200' are frequently observed in the reverse Monte Carlo simulated structure. About 40% of the Fe atoms are associated with bcc-related Fe clusters in the model structure. Trigonal prisms and Archimedean anti-prisms are also observed around the B atoms, which are mainly distributed outside the Fe clusters. Figure 2.10 shows a perspective view of the reverse Monte Carlo simulated structure with the bcc-related Fe atoms. In the model, the less dense areas show Fe(Zr)-B prism-related short range order. It has been demonstrated from both high-resolution electron microscopy and pair distribution function studies that a nanoscale phase separation exists in the amorphous $Fe_{90}Zr_7B_3$ alloy.

Changes of atomic structure on annealing

$Fe_{90}Zr_7B_3$

Because of very good soft magnetic properties, much attention has been paid to amorphous Fe-Zr-B alloys which form an α-Fe nanocrystalline structure on primary crystallization [23]. Atomic structure changes on annealing have been studied [12] in order to investigate the role of the amorphous matrix in constraining the growth of α-Fe particles in the crystallization stage [24]. Specimens were annealed at 693, 733 and 923 K for 1 h, followed by electron diffraction intensity measurements, pair distribution function analysis and reverse Monte Carlo simulation to determine changes in local atomic arrangements on annealing. At 693 and 733 K, there is no crystallization and selected area diffraction still shows halo-patterns. However, on annealing at 923 K, nanoprecipitates of α-Fe start to appear, as shown in figure 2.11, which makes the selected area diffraction pattern analysis difficult because of the appearance of bcc diffraction overlapped with the halo-pattern. Therefore, a nanodiffraction technique was used in order to analyse the matrix amorphous structure, with a probe-size of about 10 nm in diameter and a beam convergence angle as small as 3×10^{-4} rad, which does not affect the intensity profiles. Figure 2.12 shows pair distribution function profile from the nanodiffraction pattern analysis. Two small subsidiary peaks appear at radial distances of $r \approx 0.29$ and 0.32 nm on the larger r side of the first main peak at $r \approx 0.25$ nm. These peaks, marked 1, 2 and 3 in the figure, correspond to Fe-Fe, Fe-Zr and Zr-Zr correlation peaks, respectively.

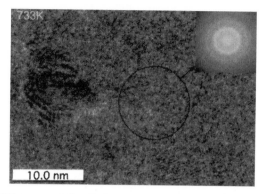

Figure 2.11 HREM image of $Fe_{90}Zr_7B_3$ annealed at 773 K for 1 h. α-Fe nanoparticles started to grow. A nanodiffraction pattern was taken from the amorphous matrix.

Partial pair distribution functions are necessary to clarify in more detail the atomic correlation distances and coordination numbers. For this purpose, reverse Monte Carlo simulations were carried out on annealed specimens using a dense random packed model as the initial structural model. Figure 2.13 shows the change of Fe atomic coordination number around Zr and the nearest-neighbour distance between Fe and Zr atoms as a function of annealing. The coordination number decreases with annealing temperature and approaches a value between 12 and 13, while the near-neighbour distance also decreases towards a value of $r \approx 0.293$ nm. A coordination of 12 and a separation of $r = 0.293$ nm correspond to averaged

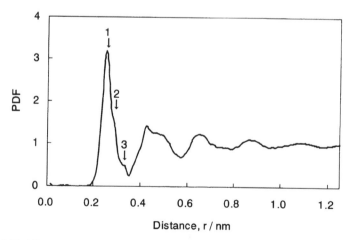

Figure 2.12. PDF profile obtained from the amorphous $Fe_{90}Zr_7B_3$ alloy annealed at 733 K. In taking the diffraction intensity, nanodiffraction was used. Three peaks indicated by 1, 2 and 3 are due to Fe-Fe, Fe-Zr and Zr-Zr correlations, respectively.

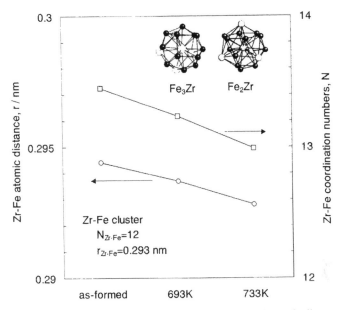

Figure 2.13. Changes of Fe coordination number around Zr and atomic distance between Zr and Fe atoms with respect to the annealing temperature. The coordination number around Zr and the atomic distance Zr-Fe reaches towards those of the averaged values for the Zr-centred coordination polyhedra in the Fe_2Zr and Fe_3Zr crystalline compounds.

values for Zr-centred coordination polyhedra in Fe_2Zr and Fe_3Zr crystalline compounds. On annealing at 923 K, Fe_2Zr and Fe_3Zr nanoprecipitates are observed by transmission electron microscopy as well as α-Fe precipitates. This structure is shown in figure 2.14, where nanocrystalline compound phases are visible as lattice images beside the α-Fe particles. The Fe-Zr correlation becomes stronger as annealing proceeds, and forms short or medium range ordered structures similar to Zr atomic structures in Fe-Zr compounds. The strong Fe-Zr correlation must be closely related to the suppression of the growth of α-Fe particles in the crystallization stage through slow diffusion of the constituent atoms. Nanoscale phase separation is expected from electron diffraction structure analysis of the as-formed state of this alloy, as shown in figure 2.10.

$La_{55}Al_{25}Ni_{20}$

Amorphous $La_{55}Al_{25}Ni_{20}$ is known to have excellent superplasticity in the stable supercooled liquid temperature range T_x [25]. In order to understand the origin of the superplastic behaviour of the alloy, it is important to study dynamic atomic structural changes in the supercooled liquid region. *In-situ* structural observations of the alloy have been performed in the supercooled

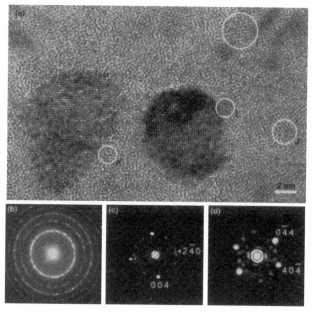

Figure 2.14. (a) HREM image of $Fe_{90}Zr_7B_3$ annealed at 923 K for 1 h. (b) SAED pattern taken from the area including the region shown in (a). Diffraction rings from randomly oriented bcc α-Fe nanoparticles are seen. (c) and (d) are Fourier-transformed patterns from areas I and J, respectively. The patterns in (c) and (d) correspond to those of Fe_3Zr [210] and Fe_2Ze [111] lattice images, respectively.

liquid region [13] by electron microscopy and diffraction, using a specimen heating stage, an image plate to record diffraction intensities and high-resolution images, and an in-column type energy filter. A series of selected area diffraction patterns and also high-resolution electron microscopy images are taken during annealing up to the crystallization temperature at a heating rate as low as 0.02 K/s.

Figure 2.15 shows a series of reduced interference functions $F(Q)$ obtained from in-situ electron diffraction during annealing in a transmission electron microscope. At 491 K crystallization of the primary phase α-La is observed by high-resolution electron microscopy and selected area diffraction. Fourier transforms of the $F(Q)$ series gives corresponding pair distribution functions $g(r)$, as shown in figure 2.16. Gradual profile changes in $g(r)$ are observed in the temperature range between 437 and 491 K, which is in the supercooled liquid range. In this temperature range, the radial distance of the main peak (mainly contributed by La-La pairs) becomes larger because of volume expansion, but becomes smaller as the temperature approaches the crystallization temperature. It is important to note from figure 2.16 that there is a large structural change in the supercooled liquid

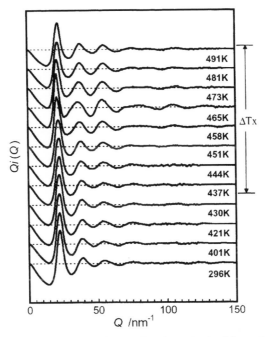

Figure 2.15. A series of reduced interference functions obtained from the *in-situ* electron diffraction study on annealing the amorphous $La_{55}Al_{25}Ni_{20}$ alloy.

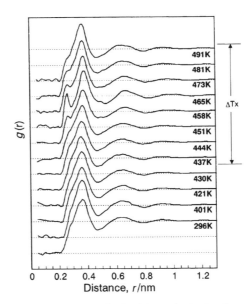

Figure 2.16. A series of atomic PDFs obtained from the *in-situ* electron diffraction study on annealing the amorphous $La_{55}Al_{25}Ni_{20}$ alloy.

Figure 2.17. HREM images taken from the as-formed (a) and the annealed (b) specimens of amorphous $La_{55}Al_{25}Ni_{20}$ alloy, respectively. The image in (b) was taken in the *in-situ* annealing experiment.

region, especially around 465 K where a subsidiary peak at $r \approx 0.26$ nm appears beside the main peak.

Figures 2.17(a) and (b) show no appreciable difference between high-resolution electron microscope images from as-formed $La_{55}Al_{25}Ni_{20}$ and after annealing in the supercooled liquid region, indicating no significant difference in atomic structure. Local fringe contrast due to medium range order is not observed frequently in either the as-formed or annealed material. Enhancement of the dense random packed nature of the alloy structure can be understood from the rapid decline in the pair distribution function profile with radial distance, as seen by comparing figures 2.5 and 2.16. This seems to indicate that the pair distribution function profiles show an atomic structure which is dense random packed and without local phase separation down to the nanometre level. The atomic structure can be understood better after reverse Monte Carlo simulation followed by Voronoi polyhedral analysis for central La, Al and Ni atoms. Figures 2.18(a)–(c) show Voronoi poly-hedral indices and fractions around La, Al and Ni atoms respectively. Lanthanum atoms are found to be coordinated by more than 10 atoms in most cases (the sum of the indices corresponds to the number of near-neighbour atoms), with a spherical shell formed by deformed fcc polyhedra with indices (03640), (03650) and others of similar type. On the other hand, aluminium and nickel atoms are coordinated by 8 and 9 atoms, mainly forming three-capped trigonal prisms and Archimedian antiprisms similar to those found in an amorphous Ti–Ni alloy [26].

The appearance of subsidiary peaks in the supercooled liquid region in figure 2.16 is judged to be caused by a change of atomic correlation on annealing. Considering the atomic radii of the constituent atoms, the subsidiary peak comes from strong atomic correlations between Al–Ni and Ni–Ni pairs. Multiple-fitting for partial correlation pair distribution function functions, especially for the first pair distribution function peak for several pair distribution function profiles. Figure 2.19 shows the results of multiple

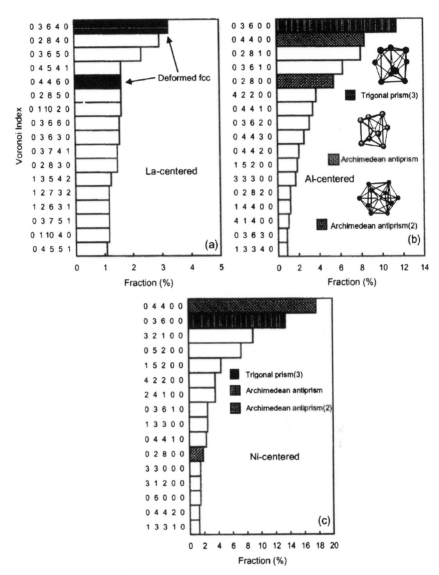

Figure 2.18. Fractions of Voronoi polyhedra around (a) La, (b) Al and (c) Ni obtained from the RMC-simulated structure for as-formed specimen of the $La_{55}Al_{25}Ni_{20}$ alloy.

fitting of the experimental pair distribution function peaks at (a) 430 K and (b) 465 K. As the annealing temperature increases, the Al-Ni (also Ni-Ni, Al-Al) correlation develops together with the La-La correlation. This shows that the local atomic structural change is towards phase separation in the supercooled liquid region. The large negative heat of mixing between aluminium and

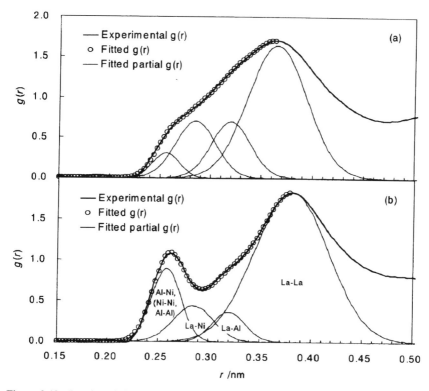

Figure 2.19. Results of the multiple fitting by Gaussian function to the first peaks of experimental PDFs obtained at temperatures of (a) 430 and (b) 465 K.

nickel atoms in addition to their high diffusivities enhance local phase separation in the supercooled liquid region. It is presumed that the evolution of the structure change originates from local atomic structure already present in the as-formed structure. As shown in figure 2.18, the atomic structure of the as-formed state resembles nanoscale phase separation with local La-La correlations forming fcc like clusters and with strong Al-Ni correlations. The strong La-La correlation in the supercooled liquid must be connected to α-La (fcc) formation during primary crystallization. The La-La correlation distance decreases when the supercooled liquid is annealed up to the crystallization temperature T_x. This is thought to be associated with a volume decrease on crystallization (the volume of the supercooled liquid is in between those of liquid and crystal phases). Superplastic behaviour must be gradually reduced after the development of atomic scale structural phase separation. Near T_g ($\sim$430 K) the random mixture of atoms is probably predominant as the atomic structure induces viscous flow and superplasticity in the alloy.

Amorphous structure of $Zr_{70}Pd_{30}$ with icosahedral atomic clusters

Icosahedral phase (I-phase) formation has been observed in a number of multicomponent Zr-based metallic glasses during primary crystallization [27–29]. Recently, the icosahedral phase has been found even in binary Zr-Pd alloys [30, 31]. Icosahedral cluster formation has been seen in various computer simulation studies of amorphous alloys in the as-formed state, although no experimental evidence has been shown. It is, therefore, interesting to investigate the local atomic structures of as-quenched and annealed amorphous alloys of Zr-based alloys which show icosahedral phase formation. Structure analysis of amorphous $Zr_{70}Pd_{30}$ has been achieved [14] in as-formed and annealed states by electron diffraction pair distribution function analysis and reverse Monte Carlo simulation, together with a high-resolution electron microscopy observation. Amorphous $Zr_{70}Pd_{30}$ alloy ribbons were obtained by single-roller melt spinning in an argon atmosphere, and halo electron diffraction patterns taken from thin specimen areas using a microelectron beam of about 100 nm in diameter and beam parallelity of 3×10^{-4} rad. Differential scanning calorimetry traces from $Zr_{70}Pd_{30}$ during annealing show two exothermic peak reactions at about 715 and 800 K. The first peak is due to icosahedral phase formation during primary crystallization, and the second peak is due to formation of the stable crystalline phase Zr_2Pd. Figure 2.20 shows a high-resolution electron microscope

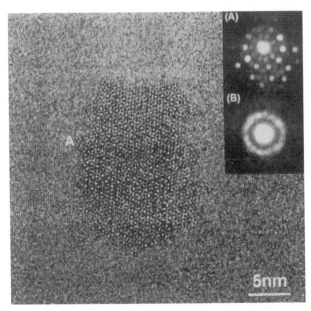

Fig. 2.20 HREM image of $Zr_{70}Pd_{30}$ annealed at 730 K showing a precipitation of I-phase.

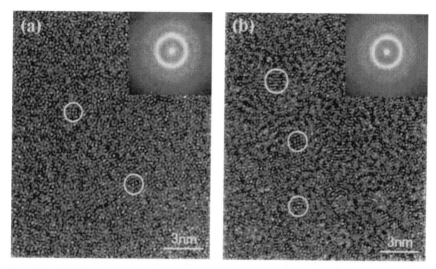

Fig. 2.21 HREM images of amorphous structures of as-formed and annealed (683 K) $Zr_{70}Pd_{30}$. MRO regions are seen in the encircled areas for example.

image representing icosahedral phase formation during annealing at 730 K. A quasi-periodic icosahedral structure is imaged in the area marked A. Nanobeam diffraction patterns with a beam size of 10 nm from the area marked A and the matrix are shown in the insets A and B. The matrix structure is mostly composed of nanosized icosahedral phase particles, although nanodiffraction patterns exhibit an amorphous-like structural feature. Figures 2.21(a) and (b) show high-resolution electron microscope images of the as-formed amorphous structure and after annealing at 633 K respectively. Corresponding electron diffraction patterns are shown in the insets. After annealing, medium range ordered regions are more extended than in the as-formed material, although structural details have not been fully identified.

Pair distribution function analyses and successive reverse Monte Carlo simulations are performed in order to understand the amorphous alloy structural changes in detail, ignoring multiple scattering with 10 nm thick specimens. Figures 2.22(a) and (b) show reduced interference functions $F(Q)$ and corresponding pair distribution functions obtained by Fourier transformation from as-quenched and annealed alloys respectively. Figure 2.22(a) shows clearly the second peak splitting with annealing. Moreover, in figure 2.22(b) the shoulder of the first peak at the large radial distance side increases with annealing, suggesting a structural change on annealing.

Modelling the amorphous structure by reverse Monte Carlo simulation uses an initial molecular dynamics structure model with 2500 atoms, Lennard-Jones atomic potentials as described above for amorphous Fe-Zr-B

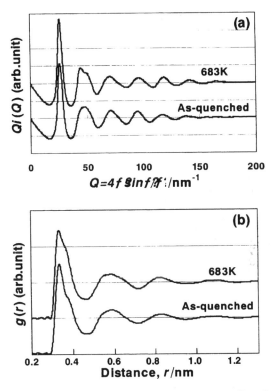

Figure 2.22. (a) Reduced interference functions and (b) PD F profiles obtained from electron diffraction intensity analysis for as-formed and annealed (683 K) $Zr_{70}Pd_{30}$.

alloys, an initial density of $7.53 \, g \cdot cm^3$ [32], and an initial slight structural relaxation. Figure 2.23 shows the results of the reverse Monte Carlo simulations. The profile fittings of the experimental pair distribution functions by the reverse Monte Carlo simulated pair distribution functions are successful. The Zr-Zr partial pair distribution function, $g_{Zr-Zr}(r)$ is sharpened with annealing, which contributes to an increase in the first peak shoulder in the pair distribution function. However, the coordination number and the interatomic distances are nearly equal before and after annealing. Figures 2.24 and 2.25 show local polyhedra centred on Pd or Zr atoms from Voroni analysis of the reverse Monte Carlo simulated structures. Voronoi polyhedra centred on Pd show predominant prism structures in both as-quenched and annealed material. The local structures surrounding Pd atoms do not change with annealing. In contrast, icosahedral structures and icosahedral-like polyhedra occupy mainly the Voronoi polyhedra centred on Zr atoms, and their fraction increases with annealing.

The presence of a large density of icosahedral and icosahedral-like clusters in the as-quenched state contributes to the precipitation of a high

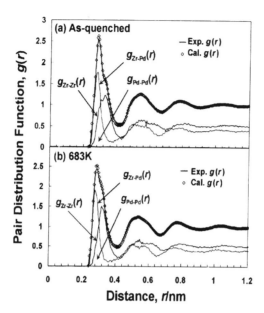

Figure 2.23. Profile fitting of experimental PDFs by RMC-simulated PDFs for (a) as-formed and (b) annealed (683 K) $Zr_{70}Pd_{30}$.

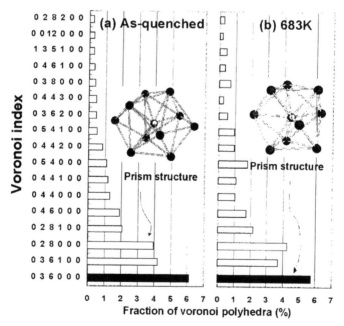

Figure 2.24. Fraction of Pd-centred Voronoi polyhedra in the (a) as-formed and (b) annealed (683 K) $Zr_{70}Pd_{30}$.

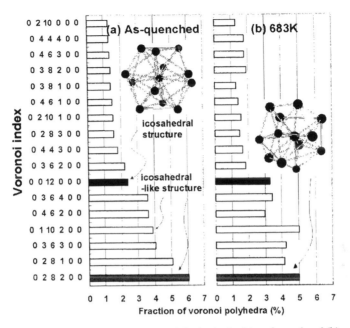

Fig. 2.25. Fraction of Zr-centred Voronoi polyhedra in the (a) as-formed and (b) annealed (683 K) $Zr_{70}Pd_{30}$.

number density of nanocrystalline icosahedral phase particles through the nucleation stage. High-resolution electron microscope observation of the as-quenched and annealed material shows no five-fold cluster medium range ordered images. Image simulation [33] shows that five-fold cluster images can hardly be imaged clearly when the clusters have sizes as small as 1.6 nm embedded in a dense randomly packed matrix, even with a small thickness electron microscope specimen of 2.8 nm.

Summary

In this chapter, modern electron microscopy and diffraction techniques are applied to determine local and average structures in amorphous alloys. The combination of these modern techniques contributes strongly to understanding local atomic structures in both as-formed and annealed amorphous alloys. Nanodiffraction is effective at determining the local organization of amorphous structures with the help of high-resolution electron microscopy, especially for extended medium range ordered structures.

The application of electron microscopy and diffraction techniques are demonstrated in medium range order structure analyses of Pd-Si alloys with different Si compositions. Selected area diffraction pattern analysis and high-resolution electron microscope observation is used to observe local atomic structural fluctuations forming nanoscale phase separation in eutectic amorphous $Pd_{82}Si_{18}$ and $Fe_{90}Zr_7B_3$ alloys. Medium range ordered regions with α-phase atomic structure are embedded in a dense random packed matrix composed mainly of metal–nonmetal prism structures. Selected area diffraction intensity analysis is also applied to structural changes during annealing of amorphous $Fe_{90}Zr_7B_3$ and $La_{55}Al_{25}Ni_{20}$ alloys. In $Fe_{90}Zr_7B_3$, a gradual change of the amorphous matrix to form Fe_2Zr or Fe_3Zr compound-like clusters is found by pair distribution function analysis and by changes in high-resolution electron microscopy images. In $La_{55}Al_{25}Ni_{20}$, a tendency to local phase separation is directly observed by *in-situ* electron diffraction in the supercooled liquid state. Icosahedral atomic cluster formation in amorphous alloys has been proposed since the 1970s and is confirmed by electron diffraction analysis in amorphous $Zr_{70}Pd_{30}$, which forms icosahedral phase precipitates during primary crystallization.

Nanostructure control on annealing amorphous alloys or bulk metallic glasses has been found to be a key to obtaining good mechanical and magnetic properties. It is essential to investigate nanostructure formation in these alloys during annealing, and modern electron microscopy and diffraction techniques can clearly play an important role.

References

[1] Stobbs W M and Smith D J 1979 *Nature* **281** 54
[2] Gaskell P H, Smith D J, Catto C J D and Cleaver J R A 1979 *Nature* **281** 465
[3] Fan G Y and Cowley J M 1985 *Ultramicroscopy* **17** 345
[4] Hirotsu Y and Akada R 1984 *Jpn. J. Appl. Phys.* **23** L479
[5] Hirotsu Y, Uehara M and Ueno M 1986 *J. Appl. Phys.* **59** 3081
[6] Anazawa K, Hirotsu Y and Inoue Y 1994 *Acta Metall. Mater.* **42** 1997
[7] Elliott S R 1990 *Physics of Amorphous Materials* Sec. 3 (Essex: Longman Sci. & Tech.)
[8] Y Hirotsu, T Ohkubo and M Matsushita 1998 *Microsc. Res. Tech.* **40** 284
[9] Matsushita M, Hirotsu Y, Anazawa K, Ohkubo T and Oikawa T 1995 *Mater. Trans. JIM* **36** 822
[10] Matsushita M, Hirotsu Y, Ohkubo T and Oikawa T 1996 *J. Electron Microsc.* **45** 105
[11] Ohkubo T, Hirotsu Y and Matsushita M 1999 *J. Electron Microsc.* **48** 1005
[12] Ohkubo T, Kai H, Makino A and Hirotsu Y 2001 *Mater. Sci. Eng.* **A 312** 274
[13] Ohkubo T, Hiroshima T, Hirotsu Y, Inoue A and Oikawa T 2000 *Mater. Trans. JIM* **41** 1385
[14] Takagi T, Ohkubo T, Hirotsu Y, Murty B S, Hono K and Shindo D 2001 *Appl. Phys. Lett.* **79** 485
[15] Buseck P, Cowley J and Eyring L (eds) 1988 *High-Resolution Transmission Electron Microscopy and Associated Techniques* (Oxford Scientific Publishers) Chap 1, p 12
[16] Nylund A 1966 *Acta Chem. Scand.* **20** 2381
[17] Canall C, Silvestri L and Gelotti G 1979 *J. Appl. Phys.* **50** 5768
[18] Hirotsu Y, Matsushita M, Hara N and Ohkubo T 1997 *NanoStruct. Mater.* **8** 1113

[19] Fukunaga T and Suzuki K 1981 *Sci. Rep. RITU* **A29** 153

[20] Duine P A, Sietsma J, Thijsse B J and Pusztai L 1994 *Phys. Rev. B* **50** 13240

[21] Kondo T and Tsumuraya K 1991 *J. Chem. Phys.* **94** 8220

[22] Gaskell P H 1985 *J. Non-Cryst. Solids* **75** 329

[23] Suzuki K, Makino A, Kataoka N, Inoue A and Masumoto T 1991 *Mater. Trans. JIM* **32** 93

[24] Zhang Y, Hono K, Inoue A, Makino A and Sakurai T 1996 *Acta Mater.* **44** 1497

[25] Inoue A 2000 *Acta Mater.* **48** 279

[26] Sietsma J and Thijsse B J 1988 *J. Non-Cryst. Solids* **101** 135

[27] Köster U, Meinhardt J, Roots S and Rüdiger A 1996 *Appl. Phys. Lett.* **69** 179

[28] Eckert J, Mattern N, Zinkevitch M and Seidel M 1998 *Mater. Trans. JIM* **39** 623

[29] Inoue A, Zhang T, Saida J, Matsushita M, Chen M W and Sakurai T 1999 *Mater. Trans. JIM* **40** 1181

[30] Murty B S, Ping D H and Hono K 2000 *Appl. Phys. Lett.* **77** 1102

[31] Saida J, Matsushita M and Inoue A 2000 *Appl. Phys. Lett.* **77** 73

[32] Waseda Y and Chen H S 1978 *Rapidly Quenched Metals III* ed. B Cantor (London: The Metals Society) pp 415–418, 480

[33] Hirotsu Y, Anazawa K and Ohkubo T 1990 *Mater. Trans. JIM* **31** 573

Chapter 3

Nanocrystalline, nanoquasicrystalline and amorphous Al and Mg alloys

Akihisa Inoue and Hisamichi Kimura

Introduction

High-strength Al- and Mg-based bulk alloys consisting of novel nanoscale non-equilibrium phases have been produced by rapid solidification and powder metallurgy techniques in Al-Ln-LTM, Al-ETM-LTM and Al-(V,Cr,Mn)-LTM (Ln = lanthanide metal, LTM = VII and VIII group transition metals, ETM = IV to VI group transition metals) alloys with high Al contents of 92–95 at% as well as in Mg-Zn-Y alloys with high Mg contents of 96–97 at%. Excellent mechanical properties have been obtained by controlling the composition, clustered atomic configuration and stability of the supercooled liquid. The non-equilibrium structures are composed of amorphous, icosahedral quasicrystalline or long periodic hexagonal phases. In particular, Al-based bulk alloys consisting of nanoscale icosahedral particles surrounded by fcc Al phase exhibit excellent mechanical properties, exceeding those for conventional high-strength type Al-based alloys. Similarly, Mg-based bulk alloys consisting of a six layered hexagonal packing matrix phase containing nanoscale cubic $Mg_{24}Y_5$ particles exhibit high strengths and large elongations which are about three times higher than those for conventional cast Mg-based alloys. This success in producing Al- and Mg-based alloys with good engineering properties by use of non-equilibrium phases is promising for future development of non-equilibrium alloys as practical materials.

Background

Since the discovery of age-hardening [1], various kinds of high-strength Al-based alloys have been produced over the subsequent nine decades. These

Al-based alloys have been developed by use of the following strengthening mechanisms: (1) solid solution strengthening, (2) precipitation strengthening, (3) grain size refinement, (4) dispersion strengthening, (5) work hardening and (6) fibre reinforcement. The use of these conventional strengthening mechanisms leads to an upper limit for the tensile strengths of Al-based alloys of 500–600 MPa at room temperature. Similar strengthening mechanisms have also been used for Mg-based alloys [2]. Conventional Mg-based alloys are usually made by various casting processes, and the strengthening has been mainly achieved by mechanisms (1), (3) and (4). The resulting strength levels lie in the range 200–300 MPa.

In order to develop new types of Al- and Mg-based alloys with much higher tensile strength, the use of a completely different strengthening mechanism seems to be essential. Systematic studies have been carried out on the development of high-strength Al- and Mg-based alloys by using non-equilibrium phase effects. In particular, much attention has been paid to non-equilibrium alloys with structures consisting of amorphous, quasicrystalline and long periodic hexagonal phases.

In 1988, our group succeeded in producing Al-based amorphous alloys with high tensile strengths exceeding 1200 MPa [3]. Subsequently [4] a homogeneous dispersion of nanoscale fcc Al particles in an amorphous matrix was found to cause a drastic increase in tensile fracture strength, up to 1560 MPa, three times higher than the strength levels of conventional high strength Al-based alloys. From the variation of tensile strength with calendar year in the historical development of high-strength Al-based alloys, a drastic increase in tensile strength is found for the recent nonequilibrium alloys including amorphous phases.

Figure 3.1 summarizes the features of the microstructure and mechanical strength of nonequilibrium Al-based alloys [5, 6]. The non-equilibrium structures of Al-based alloys are classified into six types: (1) amorphous single phase structures, (2) nanostructures consisting of Al and intermetallic compounds obtained by crystallization of an amorphous phase, (3) partially crystallized structures consisting of nanoscale fcc Al particles embedded in an amorphous matrix, (4) nanoquasicrystalline structures consisting of nanoscale quasicrystalline particles surrounded by an Al phase without grain boundaries, (5) coexisting nanogranular amorphous and Al phases, and (6) structures of a nanogranular Al phase surrounded by an amorphous network. These nonequilibrium Al-based alloys exhibit much better mechanical properties than the conventional crystalline alloys developed to date.

Mg-based bulk glassy alloys with high tensile strength were produced for the first time in the Mg-Cu-Ln and Mg-Ni-Ln systems in 1988 [7]. Tensile strengths reach 620 MPa, but the bending ductility decreases significantly because of structural relaxation caused by room temperature ageing. Because of this significant loss of ductility, investigations have moved to Mg-based solid solutions with much higher Mg contents of above 90 at%

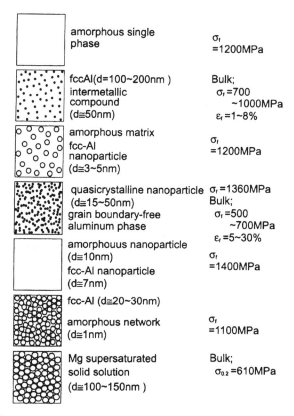

Figure 3.1. Microstructures and mechanical strength of nonequilibrium Al- and Mg-based alloys developed by the present authors.

produced by warm extrusion of gas atomized powders [8]. We have tried to develop such high-strength rapidly solidified/powder metallurgical (RS P/M) Mg-based alloys consisting of an hcp solid solution containing fine scale intermetallic compounds [8]. However, the resulting mixed phase alloys exhibit low elongations of less than 3%, and the poor ductility has prevented further development of these Mg-based alloys as structural materials. Very recently, some Mg-rich alloys in the Mg-Zn-Y system have been shown to have high strengths and ductilities, presumably because of the formation of a novel long periodic hexagonal matrix phase with much smaller grain size of 100–150 nm combined with a relatively low volume fraction of the inter-metallic compound [9].

This chapter reviews recent results on the synthesis, microstructure and mechanical properties of Al- and Mg-based alloys with non-equilibrium structural features as described above. Chapters 1 and 2 describe the fundamental thermodynamics and structure of nanocrystalline alloys, and chapters 4–9 describe the manufacture and structure of a number of different nanocrystalline metallic materials.

Al-based bulk amorphous alloys

Amorphous alloys obtained by melt spinning can be divided into two groups, metal–metal and metal–metalloid systems. Among these alloy systems, both Al-Ln-TM [3, 10] and Al-ETM-LTM [11] (Ln = lanthanide metal, TM = transition metal, ETM = IV to VI group transition metal, LTM = VII and VIII group transition metal) are the most important because of the achievement of very high tensile strengths. In addition to melt-spun Al-based amorphous alloy ribbons, melt-extracted Al-based amorphous alloy wires [12] also exhibit high tensile strengths and good bending ductility. The wire diameter is in the range 40–120 μm and the tensile strength is in the range 900–1100 MPa, nearly the same as for the melt-spun amorphous alloy ribbons. Furthermore, the use of a high-pressure die-casting method has also enabled production of Al-based bulk alloys with an amorphous surface layer [13]. The thickness of the amorphous surface layer is about 40 μm for a cylindrical ingot of 0.5 mm in diameter and increases to about 150 μm for an ingot of 5 mm in diameter. However, bulk amorphous single phase alloys have not been obtained in Al-based alloy systems by any kinds of casting method, because of their relatively low glass-forming ability.

Amorphous alloys with high Al concentrations of about 85 at% in Al-Ln-TM systems exhibit a glass transition, followed by a supercooled liquid region below the crystallization temperature, though the supercooled liquid region is rather narrow and its temperature interval is less than 30 K [14]. In the supercooled liquid region, significant viscous flow is obtained over an appropriate strain rate range. It is expected that bulk amorphous Al-based alloys can be produced by extrusion of atomized amorphous powders in the supercooled liquid state. By using gas atomization, amorphous alloy powders without any trace of crystallinity are produced in particles with sizes below 25 μm [15]. When an $Al_{85}Ni_{10}Ce_5$ amorphous powder in the size fraction below 25 μm is extruded at temperatures between 443 and 493 K, at an extrusion velocity of 0.1 mm/s and at extrusion ratios of 2.25 and 2.50, cylindrical bulk alloys with diameters of 10, 15 and 20 mm are produced, as shown in figure 3.2 [16]. X-ray diffraction patterns from the bulk alloys consist of a halo peak due to the formation of a single amorphous phase. This is believed to be the first synthesis of Al-based bulk alloys with a single amorphous phase by extrusion. This success in synthesizing Al-based consolidated bulk alloys is attributed to more precise control of extrusion conditions, based on more detailed information on the viscous flow behaviour of supercooled liquids for typical bulk amorphous alloys such as Zr- and Pd-based ones [17, 18]. Within

Figure 3.2. Outer shape and morphology of bulk amorphous $Al_{85}Ni_{10}Ce_5$ alloys produced by warm extrusion of atomized amorphous powders.

several years, further extension of the present consolidation technique may enable us to produce Al-based bulk amorphous alloys with high strengths and good ductilities, nearly the same as those for the corresponding melt-spun amorphous alloy ribbons.

Al-based partially crystallized alloys

It is well known [19] that the phase transformation from a highly supercooled liquid or amorphous phase has the following unique features: (1) homogeneous nucleation, (2) high nucleation frequency, (3) low growth rate, (4) high concentration gradient of solute elements at the liquid/solid interface resulting from low atomic diffusivity, (5) formation of metastable phases with new compositions by controlled redistribution of solute elements, (6) formation of a residual amorphous phase with high solute concentration, (7) defect-free nanocrystalline particles with low residual strain, (8) highly dense-packed atomic configurations at the liquid/solid interface, (9) nanoscale interparticle spacings, and (10) size and shape effects of nanoscale spherical particles. By utilizing these features, a new nanoscale mixed phase alloy containing a remaining amorphous phase can be synthesized for Al-based and other alloys.

In general, Al-rich amorphous alloys with high Al concentrations of above 88 at% crystallize through two stages, in which the first-stage exothermic reaction is due to the precipitation of fcc Al, and the second-stage exothermic peak results from the decomposition of the remaining amorphous phase to form an intermetallic compound [19]. When the cooling rate of an Al-rich alloys is controlled, a nanoscale mixed structure consisting of fcc Al particles with a size of 3–5 nm embedded in an amorphous matrix can be produced in the melt-spun ribbons [20]. The volume fraction V_f of the fcc Al phase can be controlled by changing the surface speed of wheel

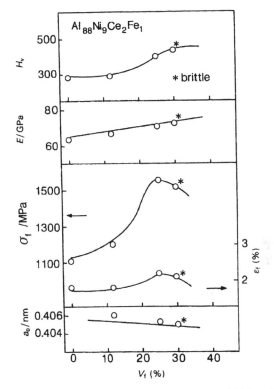

Figure 3.3. Tensile fracture strength σ_f, Young's modulus E and Vickers hardness H_V as a function of volume fraction V_f of Al phase for melt-spun $Al_{88}Ce_2Ni_9Fe_1$ alloy ribbons prepared at different cooling rates.

and can be evaluated by the change in the exothermic heat effect value due to crystallization. The tensile fracture strength σ_f increases from 1100 MPa at $V_f = 0\%$ to 1560 MPa at $V_f = 25\%$, accompanying increases in Vickers hardness H_V from 280 to 400 and Young's modulus E from 63 to 71 GPa, as shown in figure 3.3. A significant decrease in fracture strength with further increase in volume fraction is caused by embrittlement of the remaining amorphous phase associated with structural relaxation and enrichment in solute elements.

A similar nanoscale mixed structure is also formed by annealing in the first-stage exothermic reaction range and the mixed phase alloys again exhibit high tensile strengths exceeding 1400 MPa, which is about 1.4 times higher than that for the corresponding amorphous single phase alloys [21]. The increase in fracture strength caused by the dispersion of nanoscale fcc Al particles in an amorphous phase has been thought [19] to result from a combination of the following three effects: (1) defect-free nanoscale fcc Al particles, because they are too fine to contain dislocations, (2) amorphous/Al particle interfaces with highly dense-packed atomic configurations, no excess vacancies and a low interfacial energy, and (3) Al nanoscale particles, smaller than the width of inhomogeneous shear bands, so they act as effective barriers against shear deformation of the amorphous matrix. However, the tensile strength is too high to obtain a consolidated bulk alloy by warm extrusion in a relatively low temperature range where the nanoscale mixed structure is maintained. Even at the present time, we are not able easily to synthesize a bulk nanostructured alloy with high tensile strength and good ductility.

Al-based bulk nanocrystalline alloys

When the extrusion temperature is raised to the second-stage exothermic reaction range, a fully dense nanocrystalline alloy can be produced, consisting of fine intermetallic compounds with sizes of about 50 nm embedded in an Al matrix with a grain size of 100–200 nm [22]. The relation between tensile strength and rotating-beam bending fatigue strength for nanocrystalline and conventional Al-based alloys is shown in figure 3.4. The nanocrystalline alloys exhibit good combinations of high tensile strength above 900 MPa and high fatigue strength of 300–350 MPa, much superior to conventional

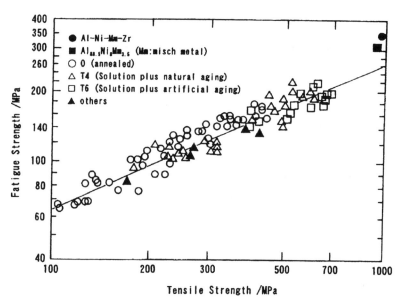

Figure 3.4. Relation between tensile strength and fatigue strength for nanocrystalline Al-Ni-Mm (Mm: misch metal) base alloys prepared by the powder metallurgy technique. Data from conventional Al-based alloys are also shown for comparison.

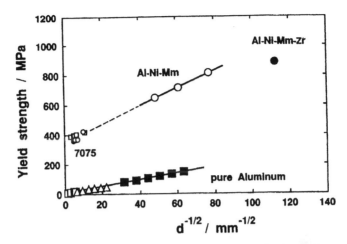

Figure 3.5. Relation between the yield strength $\sigma_{0.2}$ and the grain size of Al matrix phase $d^{-1/2}$ for nanocrystalline Al-Ni-Mm based alloys prepared by the powder metallurgy technique. Data from conventional Al-based alloys are also shown for comparison.

Al-based alloys [23] as well the new Al-based alloys [24] developed by rapid solidification and powder metallurgy. Figure 3.5 shows the relation between the yield strength $\sigma_{0.2}$ and the grain size of the fcc Al matrix for the nanocrystalline Al-based alloys, together with data for conventional crystalline Al-based alloys. The high tensile strength has been explained as resulting from a combination of two mechanisms, dispersion strengthening and grain size refinement. This combination of two strengthening mechanisms is attributable to the formation of a unique mixed structure in which the intermetallic compounds are dispersed homogeneously within the Al grains and on the grain boundaries. The dispersion density of the compounds is not high enough to suppress the movement of all dislocations and a significant number are hindered at the Al grain boundaries. Figure 3.5 also shows that the increase in the yield strength by dispersion strengthening and grain size refinement are approximately 400 and 390 MPa respectively. When dispersion strengthening is attributable to an Orowan mechanism [25], the increase in the yield strength is given by

$$\Delta\tau = 0.81Gb/2\pi(1-\nu)^{1/2}\ln(2r_s/r_o)/(\lambda_s - 2r_s) \qquad (3.1)$$

where G is the Al shear modulus (27 GPa), b is the Al Burger's vector (2.86×10^{-10} m), ν is the Al Poisson's ratio 0.345, r_s is the particle radius, r_o is the core dislocation radius for Al (2.86×10^{-10} m), λ_s is $1.25r(2\pi/3f)^{1/2}$ and f is the volume fraction of dispersoids. Values of r and f for the present alloy are measured to be 2.5×10^{-8} m and 0.34 respectively [26] and $\Delta\tau$ is calculated to be 129 MPa so the resulting increase of yield stress $\Delta\sigma$ is 395 MPa.

The increase in yield strength from grain size refinement is given by the Hall–Petch relation [27, 28]

$$\sigma_{0.2} = \sigma_o + kd^{-n} \tag{3.2}$$

where σ_o is the yield strength of a single grain due to all strengthening mechanisms excluding grain boundary contributions [29], k is a constant related to how effective the grain boundaries are in increasing yield strength, d is the grain or subgrain size and n is an exponent in the range 0.5–1 and assumed to be 0.5 [29]. In the present alloy, σ_o includes solid solution strengthening and strain strengthening up to 0.2% elongation as well as dispersion strengthening. The k value has been reported to be 75 MPa $\mu m^{1/2}$ for commercial pure Al [30], 120 MPa $\mu m^{1/2}$ for optimally aged 7075 alloy [31] and 220 MPa $\mu m^{1/2}$ for underaged 7091 alloy [32]. In the calculation, the k value was assumed to be 135 MPa $\mu m^{1/2}$. The resulting kd^{-n} value is 426 MPa for $d = 100$ nm and 302 MPa for $d = 200$ nm. Consequently, the yield strength $\sigma_{0.2}$ evaluated by summation of the dispersion strengthening and grain size refinement effects is predicted to be 821 MPa for $d = 100$ nm and 697 MPa for $d = 200$ nm. These calculated $\sigma_{0.2}$ values agree with the experimental results shown in figure 3.5, confirming the operation of dispersion and grain size strengthening mechanisms.

The nanocrystalline structure has been found [33] to lead to high-strain-rate superplasticity as evidenced from a high-strain-rate sensitivity exponent (m-value) of 0.5 as well as a large elongation reaching about 670% at a high strain rate of $1.0\,s^{-1}$. These nanocrystalline Al-based alloys have already been commercialized under the commercial name of GIGAS [34] because they exhibit high tensile strengths of about 1 GPa. The GIGAS alloys also have higher specific strength and higher specific Young's modulus which are superior to conventional Al-based and Ti-based alloys. These advantages have already enabled applications in fields such as rapid-repeating machinery parts requiring high specific strength and high specific modulus, main construction parts in robots and high elevated temperature die casting materials.

Al-based bulk nanoquasicrystalline alloys

It is known that an icosahedral phase is formed in rapidly solidified Al-TM base alloys containing Mn [35], Cr [36] and V [37, 38] as the TM element. The Al-TM base icosahedral structure has been presumed [39] to consist of Mackay icosahedral clusters containing 55 atoms which are arranged through glue atoms into a three-dimensional quasiperiodic structure. The icosahedral phase can contain dislocations [40, 41]. However, it is very difficult for dislocations to move in the icosahedral phase because their movement destroys the quasiperiodic lattice at both room and elevated temperatures [40]. A large single icosahedral $Al_{70}Pd_{10}Mn_{20}$ quasicrystal has been prepared [42] by the Chochralski method and the mechanical properties measured as a function of quasicrystal orientation and temperature [43]. The stoichiometric quasicrystal has a high Young's modulus of 200 GPa, a high rigidity modulus of 72 GPa and a high Vickers hardness of about 750 at room temperature, even though it is extremely brittle [43]. It also exhibits a high elevated temperature compressive strength of about 650 MPa at 1000 K, with significant ductility [43]. By utilizing the high Young's modulus, high hardness and high elevated temperature strength, an icosa-hedral-based Al alloy with high strength is capable of being developed. The key point in the development is the possibility of synthesizing an icosahedral-based alloy with good ductility.

High strength and good ductility for icosahedral-based alloys are expected for a nanostructured state where icosahedral phase particles have spherical morphology, small grain sizes and are homogeneously dispersed throughout the Al matrix. A nanostructured state is expected to be obtained by increasing the stability of the supercooled liquid against transition to a crystalline phase. Consequently, significant attention has been paid to Al-Mn-Ln and Al-Cr-Ln systems, because Al-Mn and Al-Cr alloys are well-known icosahedral-forming system and Al-Ln binary alloys are good glass-formers. Figure 3.6 shows the compositional dependence of structure, ductility and tensile fracture strength of melt-spun Al-Mn-Ce alloys [44]. A mixed structure consisting of icosahedral and aluminium phases is formed in the composition range above 92 at% Al. These mixed phase alloys have good bending ductilities and high fracture strengths, reaching 1320 MPa.

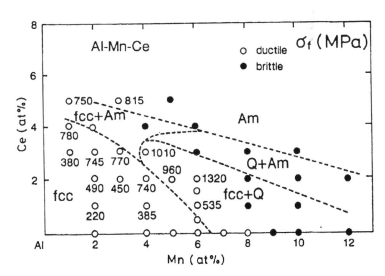

Figure 3.6. Compositional dependence of structure, ductility and fracture strength of melt-spun Al-Mn-Ce alloys.

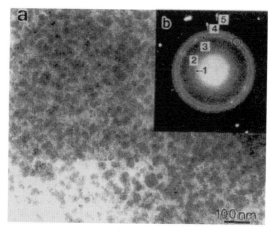

Figure 3.7. Bright-field electron micrograph and selected-area electron diffraction pattern of melt-spun $Al_{94.5}Cr_3Ce_1Co_{1.5}$ alloy. The reflection rings of 1, 2, 3, 4 and 5 are identified to be (111100), (211111), (221001), (322101) and (332002), respectively.

High-strength $Al_{92}Mn_6Ce_2$ consists of spherical icosahedral particles with a size of about 50 nm surrounded by the fcc Al phase with a thickness of 5–10 nm. Halo-like reflection rings result from the icosahedral particles, which have random orientation because of the primary precipitation of the icosahedral phase, followed by the solidification of the aluminium phase from the remaining liquid. The volume fraction of the icosahedral phase is as high as 60–70%. A similar nanoscale mixed structure is also formed in rapidly solidified Al-Cr-Ce-Co alloys [45]. As exemplified for $Al_{94.5}Cr_3Ce_1Co_{1.5}$ in figure 3.7, icosahedral particles with a size of 20–40 nm are dispersed homogeneously in the aluminium matrix, without high-angle grain boundaries. The icosahedral-based alloy also exhibits a high fracture strength of 1340 MPa.

In general, high fracture strengths exceeding 1000 MPa combined with high hardnesses above 400 are also obtained for rapidly solidified $Al_{93.5}Cr_3Ce_1Co_{1.5}TM_1$ alloys containing various transition metals in addition to Cr and Co [46]. The icosahedral-based alloys also have good cold deformability and can be cold rolled up to significant thickness reduction ratios, above 70% [47]. Even in the heavily deformed alloys, no appreciable cracks are observed. It has been confirmed that the cold-rolled alloy has a much finer mixed structure consisting of an icosahedral phase with a size of 5–10 nm in coexistent with nanoscale fcc Al phase.

Figure 3.8 illustrates the change in the mixed structure of icosahedral and Al phases with solidification mode. High tensile strength combined with good ductility is obtained only for mixed phase alloys prepared through the unique solidification process in which the nanoquasicrystalline phase precipitates as a primary phase followed by the solidification of Al

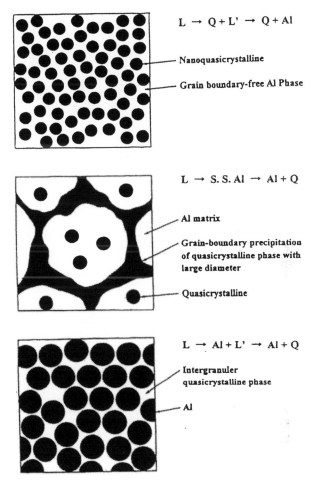

Figure 3.8. Illustration of various formation processes of mixed icosahedral and Al phases in melt-spun Al-based alloys.

from the remaining liquid. High tensile strength and good ductility cannot be obtained when the alloys have other types of solidification process, when a primary Al solid solution is followed by precipitation of the icosahedral phase between the Al grains and along the grain boundaries or followed by solidification of an intergranular quasicrystalline phase from the remaining liquid.

In addition to melt-spun ribbons, the same mixed structure of nanoscale icosahedral phase surrounded by an Al matrix without grain boundaries is also formed in atomized powders of Al-Mn-Ln and Al-Cr-Ln-TM systems as well as Al-Mn-TM and Al-Cr-TM systems without lanthanide elements, as exemplified for $Al_{93}Mn_5Co_2$ in figure 3.9 [48]. When the atomized

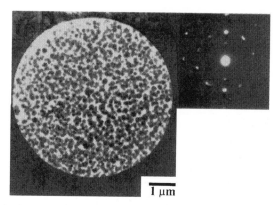

Figure 3.9. Bright-field electron micrograph and selected-area electron diffraction pattern of an atomized $Al_{93}Mn_5Co_2$ alloy powder with a particle size fraction below $45\,\mu m$.

powders are extruded in the temperature range below 673 K, lower than the decomposition temperature of the icosahedral phase, fully dense bulk icosahedral-based alloys can be obtained with the same mixed structure as the atomized powders. Figure 3.10 shows the relation between tensile strength and plastic elongation for the extruded bulk icosahedral-based

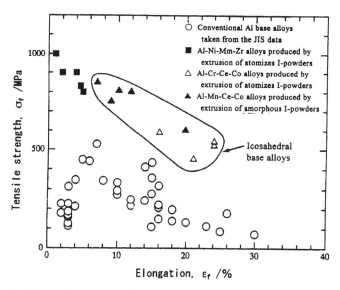

Figure 3.10. Relation between tensile strength and plastic elongation for the extruded bulk icosahedral base alloys. Data for conventional Al-based alloys are also shown for comparison.

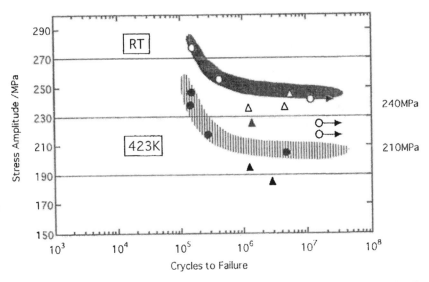

Figure 3.11. Relation between maximum applied stress and cycles up to failure for the extruded bulk icosahedral base $Al_{94}Cr_{2.5}Co_{1.5}Mm_1Zr_1$ alloys.

alloys [48, 49] in comparison with data for conventional Al-based alloys obtained from the JIS handbook. The bulk icosahedral-based alloys have high fracture strengths of 500–850 MPa combined with large elongations of 5–25%, which are much superior to those for conventional Al-based alloys.

The bulk icosahedral-based alloys can be divided into the following two groups: (1) high-strength Al-Mn(Cr) alloys containing lanthanide elements, and (2) high ductility Al-Mn(Cr)-TM alloys without lanthanide elements. The Al-Cr and Al-Mn-based alloys containing lanthanide elements exhibit rather high elevated temperature strengths of 350 MPa at 473 K and 200 MPa at 573 K, in addition to high Young's moduli of about 100 GPa at room temperature. Bulk icosahedral $Al_{94}Cr_{2.5}Co_{1.5}Mm_1Zr_1$ exhibits high fatigue strength values of 240 MPa at room temperature and 210 MPa at 423 K, much higher than 175 and 125 MPa respectively for conventional Al-based alloys in the Al-Si-Fe systems, as shown in figure 3.11. Bulk icosa-hedral-based alloys without lanthanide elements in the $Al_{93-95}(Cr,Mn,Ni)_{5-7}$ and $Al_{93-95}(Cr,Mn,Cu)_{5-7}$ systems exhibit large elongations reaching 30% and high impact fracture energies reaching 160 kJ/m^2, combined with high tensile strengths of 500–800 MPa [50]. All the values of tensile strength, elongation and impact fracture energy are much higher than those for the conventional 7075-T6 alloy.

A near-single icosahedral phase has recently been found at a new com-position of $Al_{84.2}Fe_{7.0}Cr_{6.3}Ti_{2.5}$ in the melt-spun state, as shown in figure 3.12

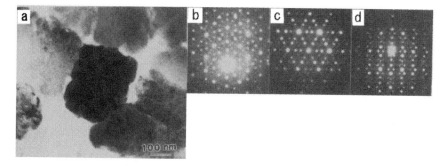

Figure 3.12. Bright-field electron micrograph and selected-area electron diffraction pattern of a melt-spun $Al_{84.2}Fe_{7.0}Cr_{6.3}Ti_{2.5}$ alloy.

[51]. This is believed to be the first evidence of the formation of an icosahedral phase in Al-Fe-based alloys. Based on previous data [52] that all the solute elements Fe, Cr and Ti have very low diffusivities in fcc Al, the icosahedral-based alloys are expected to exhibit high elevated temperature strengths. Even for an $Al_{93}Fe_3Cr_2Ti_2$ alloy with a high Al content, a

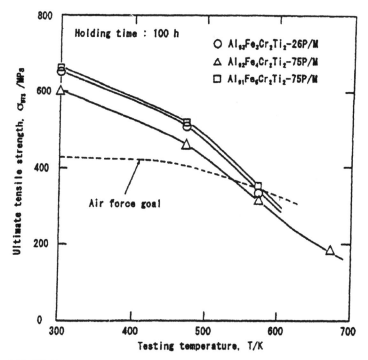

Figure 3.13. Yield strength ($\sigma_{0.2}$) and tensile fracture strength (σ_f) as a function of testing temperature for extruded bulk icosahedral base Al-Fe-Cr-Ti alloys. Data for other Al-based alloys with high elevated temperature strength are also shown for comparison.

Table 3.1. Alloy system, microstructures and mechanical properties for quasicrystalline-based bulk Al alloys.

Type	Alloy system	Structure	Mechanical properties
High-strength alloy	Al-Cr-Ce-M Al-Mn-Ce	$Al + Q$	$\sigma_f = 600\text{--}800\,\text{MPa}$ $\varepsilon_P = 5\text{--}10\%$
High-ductility alloy	Al-Mn-Cu-M Al-Cr-Cu-M	$Al + Q$	$\sigma_f = 500\text{--}600\,\text{MPa}$ $\varepsilon_P = 12\text{--}30\%$
High-elevated temperature strength alloy	Al-Fe-Cr-Ti	$Al + Q + Al_{23}Ti_9$	$\sigma_f = 350\,\text{MPa}$ at 573 K

similar mixed structure mainly consisting of icosahedral and aluminium phases is formed in atomized powders with a particle size below 125 μm [51]. The resulting bulk alloys prepared by extrusion of the atomized powders also consist mainly of icosahedral and Al phases. Figure 3.13 shows the elevated temperature strength of the bulk icosahedral-based alloys, in comparison with data for the air-force goal and conventional high-strength Al alloys [51]. The new Al-Fe-Cr-Ti alloys exhibit excellent elevated temperature strengths of 400–460 MPa at 473 K and 350–360 MPa at 573 K which exceed the air-force goal. It has further been confirmed that the high elevated temperature strengths of extruded $Al_{93}Cr_3Fe_2Ti_2$ is maintained even after annealing for 1000 h at 573 K. The annealed alloy maintains a mixed structure of fine scale icosahedral particles with a size less than 400 nm and fine scale Al grains with a size of about 500 nm.

Table 3.1 summarizes the alloy systems, microstructures and mechanical properties of the quasicrystalline-based Al alloys. The icosahedral-based Al alloys can be classified into three groups: (1) high strength alloys in Al-Mn-Ln and Al-Cr-Ln-TM systems, (2) high ductility alloys in Al-Mn-Cu-TM and Al-Cr-Cu-TM systems, and (3) high-elevated temperature strength alloys in the Al-Fe-Cr-Ti system. When the mechanical properties are compared with those for 7075-T6, the present nanoquasicrystalline alloys exhibit much better combinations of almost all properties including tensile strength, elevated temperature strength, Charpy impact fracture energy, cold workability, elongation and Young's modulus, except thermal expansion, as illustrated in figure 3.14. These excellent characteristics are quite attractive for future application as a new type of Al-based alloys with high fracture strength, high ductility and high elevated temperature strength.

Al-based nanogranular amorphous alloys

We have examined the structure and mechanical properties of melt-spun Al-V and Al-Ti based alloys containing solute elements belonging from groups IV

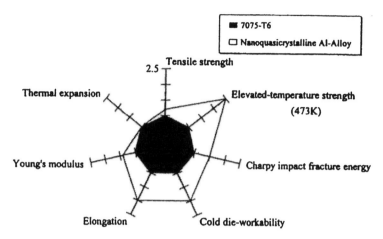

Figure 3.14. Comparison of mechanical properties of the icosahedral base bulk alloys with those for the commercial 7075-T6 alloy.

and V in the periodic table. The $Al_{94}V_4Fe_2$ alloy consists of nanogranular amorphous grains with a size of about 10 nm coexisting with an Al phase with a grain size of about 7 nm [53]. A similar nanogranular amorphous structure is also formed in melt-spun $Al_{95}Ti_3Fe_2$ [54]. These nanogranular mixed structure alloys in melt-spun ribbon form have good bending ductility and exhibit high strengths of 1390 MPa for $Al_{94}V_4Fe_2$ and 1200 MPa for $Al_{95}Ti_3Fe_2$. Table 3.2 summarizes the structures of melt-spun Al-Ln-TM and Al-ETM-LTM alloys containing high Al contents from 93 to 95 at%. There is a systematic change in the structures:

$$Al + compound \rightarrow Al + icosahedral \; quasicrystal$$

$$\rightarrow Al + amorphous \rightarrow amorphous$$

with decreasing group number of the transition metal alloying elements. Such a systematic change in the rapidly quenched structure is presumably because the decrease in group number causes an increase in the degree of satisfaction of the rules for achieving a stabilised supercooled liquid, with increases in the atomic size ratios and negative heats of mixing among the

Table 3.2. Structures of melt-spun Al-Ln-TM and Al-ETM-LTM alloys containing high Al contents of 93 to 95 at%.

Element	Lanthanide	ETM			LTM
	Ln	Ti	V	Cr-Mn	Fe-Co-Ni-Cu
Structure	Am	Al + Am	Al + Am/AL + Q	Al + Q	Al + Comp.

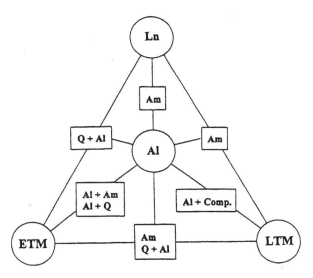

Figure 3.15. Structures of melt-spun Al-based binary and ternary alloys.

constituent elements of the alloy. The structures of the melt-spun Al-based binary and ternary alloys are summarized in figure 3.15. Easy formation of nonperiodic amorphous and nanoquasicrystalline structures are found in Al-rich composition ranges above about 92 at% Al of Al-Ln and Al-EM binary and Al-Ln-EM, Al-Ln-LM and Al-EM-LM ternary systems by melt spinning. These nonperiodic Al-based alloys form a new type of high-strength engineering materials.

Mg-based bulk amorphous alloys

The formation of an amorphous phase in Mg-based alloys by rapid solidification was reported for Mg-Ca [55] and Mg-Zn [56] systems in the 1970s. In 1988, we produced for first time Mg-based glassy Mg-Cu-Ln and Mg-Ni-Ln alloys with a large supercooled liquid region before crystallization [57]. As examples, figure 3.16 shows the composition range in which a glassy phase is formed in Mg-Ni-R and Mg-Cu-R (rare earth R = Y, Nd, La, Ce) ternary systems by melt spinning [58]. Amorphous alloys are formed over very wide composition ranges at less than about 88 at% Mg. By choosing Mg-based alloys containing 60–80 at% Mg, bulk amorphous alloys with diameters up to 8 mm can be produced by the copper mould casting and high-pressure die casting methods [59]. Very recently, bulk amorphous Mg-based alloys have been made with diameters up to 12 mm in the Mg-Cu-Y-Ag-Pd system, as exemplified in figure 3.17 [60]. Bulk amorphous Mg-based alloys with diameters of 8, 10 and 12 mm can be produced by water quenching the

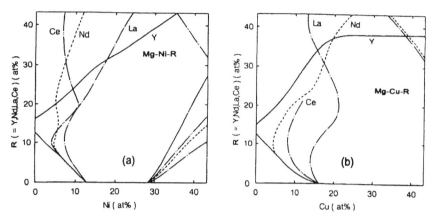

Figure 3.16. Composition ranges in which an amorphous phase is formed in Mg-Ni-R and Mg-Cu-R (R = Y, Nd, La or Ce) systems by melt spinning.

alloy melt in a carbon steel tube. The simultaneous addition of noble metals Ag and Pd with significantly different atomic sizes to Mg-Cu-Y ternary alloys is effective for further increase in glass-forming ability through an increase in reduced glass transition temperature T_g/T_m, caused by a decrease in melting temperature T_m. The bulk amorphous Mg-based alloys exhibit high mechanical strengths. Figure 3.18 shows the tensile stress-elongation curves of bulk amorphous $Mg_{80}Cu_{10}Y_{10}$ alloy sheet with a thickness of 2 mm. The tensile fracture strength is as high as 620 MPa at room temperature and decreases with increasing testing temperature, with a significant increase in elongation [59]. The tensile fracture strength of 620 MPa is attractive enough to be used for practical uses as a structural material. However, the bulk amorphous alloy becomes brittle after room temperature ageing for about 6 months, presumably because of structural relaxation [61]. As a result, Mg-based bulk amorphous alloys have not yet been used as structural materials.

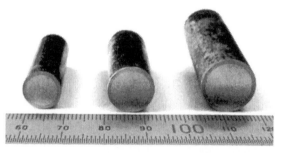

Figure 3.17. Outer shape of bulk glassy $Mg_{60}Cu_{20}Y_{10}Ag_5Pd_5$ alloy rods with diameters of 8, 10 and 12 mm.

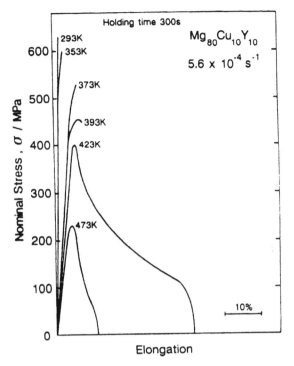

Figure 3.18. Tensile stress-elongation curves at different testing temperatures for cast bulk glassy $Mg_{80}Cu_{10}Y_{10}$ alloy sheet with a thickness of 2 mm.

Bulk hcp Mg-Ln-TM and Mg-Ln-Ln alloys

The compositional dependence of microstructure, Vickers hardness and bend ductility has been examined in melt-spun Mg-Ln-TM and Mg-Y-Mm alloys, with the aim of finding an alloy composition with elevated hardness and good ductility. The alloys show good bend ductility and high hardness. Consolidated bulk alloys can be produced by warm extrusion of high pressure gas atomized alloy powders. The overall process consists of producing alloy powders by high-pressure gas atomization, followed by collection, sieving, pre-compaction, sealing into Al cans and then extrusion at a ratio of 10:1 in the temperature range 573–723 K. Figure 3.19 shows tensile stress-elongation curves at room temperature for bulk $Mg_{95.5}Mm_2Y_{2.5}$ alloys produced at different extrusion temperatures ranging from 573 to 698 K [8]. The tensile yield strength exceeds 500 MPa in the extrusion temperature range of 573–648 K, but the elongation is less than 1.0%. The low elongation is a disadvantage for bulk hcp Mg-based alloys in the Mg-Mm-Y and Mg-Ln-TM systems. On the contrary, Mg-based solid solution

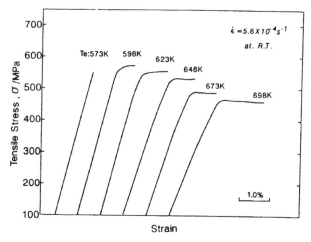

Figure 3.19. Tensile stress-elongation curves of rapidly solidified powder metallurgy (RS P/M) $Mg_{95.5}Mm_2Y_{2.5}$ alloys produced at various extrusion temperatures.

alloys exhibit very high elevated temperature strengths. Figure 3.20 shows that for $Mg_{95}Mm_3Zn_2$ and $Mg_{95.5}Mm_2Y_{2.5}$ alloys, high yield strengths above 400 MPa are maintained over the temperature range up to 523 K, with strengths 2 to 3 times higher than for the commercial WE54A-T6

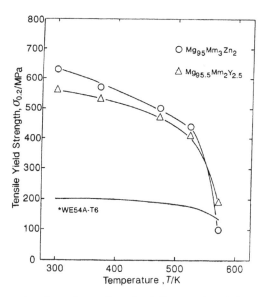

Figure 3.20. Temperature dependence of tensile yield strength and elongation for RS P/M $Mg_{95}Mm_3Zn_2$ and $Mg_{95.5}Mm_2Y_{2.5}$ alloys. Data for conventional WE54A-T6 alloys are also shown for comparison.

alloy in a wide temperature range below 550 K [2]. Similar high elevated temperature strength is obtained for extruded bulk $Mg_{95}Ca_{2.5}Zn_{2.5}$ and $Mg_{96.5}Ca_1Zn_{2.5}$ alloys, though the corresponding elongations are lower than 3% [62].

Strengthening mechanism in bulk hcp Mg-based alloys

Figure 3.21 shows the relation between tensile yield strength and grain size $d_g^{-1/2}$ for the Mg-based solid solution alloys. There is a good linear relationship, within some experimental scatter. This indicates that the high strength is partly due to grain size refinement of the hcp Mg matrix. As exemplified for the $Mg_{95.5}Y_2Mm_{2.5}$ alloy in figure 3.22, consolidated bulk alloys made from atomized powders have two kinds of structure: (1) hcp solid solution single phase, and (2) mixed structure consisting of fine scale intermetallic compounds embedded in an hcp Mg matrix. It is therefore presumed that the scatter in the data in figure 3.21 is due to the existence of fine scale intermetallic compounds. In order to reduce the degree of scattering, the tensile yield strengths are plotted in figure 3.23 as a function of $V_f^{3/2}r^{-1}$ which is known as the Fisher–Hart–Pry parameter [63]. V_f is the volume fraction of the second phase and r is the interparticle spacing. A much better linear

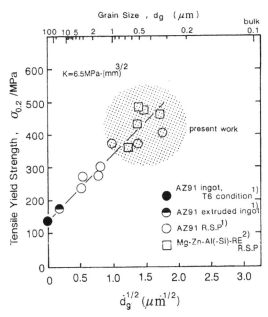

Figure 3.21. Relation between tensile yield strength and grain size $d_g^{-1/2}$ for various rapidly solidified Mg-based alloys.

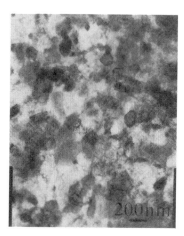

Figure 3.22. Bright-field transmission electron micrograph of RS P/M $Mg_{95.5}Y_2Mm_{2.5}$ alloy.

relationship is obtained, indicating that the high strength is due to a combination of grain size refinement and dispersion hardening. The high elevated temperature strength shown in figure 3.20 is concluded to result mainly from dispersion hardening.

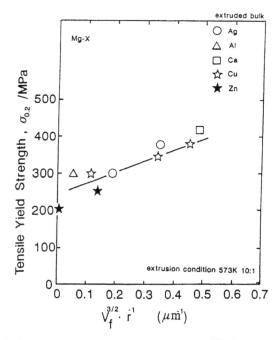

Figure 3.23. Relation between tensile yield strength and $V_f^{3/2}r^{-1}$ for RS P/M Mg-based alloys.

Long period hcp alloys

The tensile yield strength $\sigma_{0.2}$ and elongation δ of the bulk hcp Mg-based alloys are shown in figure 3.24 [9], including data from rapidly solidified powder metallurgy (RS P/M) AZ91 and ZK61 alloys, as well as JIS data from ingot metallurgy (IM) ageing Mg-based alloys [2]. Based on previous knowledge that low elongations below 3% prevent practical uses of Mg-based alloys, we tried to develop a completely new Mg-based alloy with much higher yield strengths and ductilities, corresponding to the target region shown in figure 3.24. New ternary Mg-based systems have been developed which satisfy the following three criteria: (1) multicomponent systems consisting of more than three elements, (2) significant atomic size ratios above about 12% among the three elements, and (3) suitable negative heats of mixing among the elements. These three criteria are very effective for stabilising metallic supercooled liquids [64, 65]. The stabilisation is expected to be associated with the formation of a highly non-equilibrium solid solution. The Mg-Al-Zn, Mg-Ca-Al and Mg-Zn-Y systems satisfy all three criteria. Microstructure, Vickers hardness and bend ductility of the melt-spun alloys have been investigated, after annealing for 1.2 ks at 573 and 673 K corresponding to thermal histories during warm consolidation. The aim is to determine an appropriate composition with high hardness while maintaining good bend ductility [66]. The composition dependence of hardness and bend ductility were examined systematically in the annealed

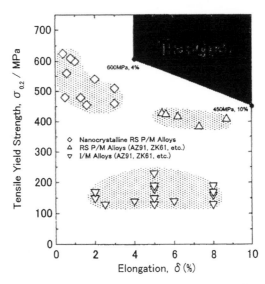

Figure 3.24. Target values of tensile yield strength ($\sigma_{0.2}$) and elongation (δ) for the present study. Data for RS P/M AZ91 and ZK61 alloys, and conventional IM-ageing Mg-based alloys are also shown for comparison.

state for melt-spun Mg-Ca-Al and Mg-Zn-Y alloys. Higher hardness and maintenance of good bend ductility was obtained for $Mg_{85}Ca_5Al_{10}$ and $Mg_{97}Zn_1Y_2$ alloys. Alloy powders of these two alloys with an average size of about $25\,\mu m$ were produced by high pressure gas atomization. The atomized powders were subsequently collected, sieved, pre-compacted, sealed into Al cans and then extruded with an extrusion ratio of $10:1$ in the extrusion temperature range $573–723\,K$. The production procedures were undertaken under a well-controlled argon gas atmosphere with an oxygen content below $0.5\,ppm$.

Atomized $Mg_{85}Ca_5Al_{10}$ powder below $25\,\mu m$ in size consists of hcp Mg, Al_2Ca and unknown non-equilibrium phases. The structure changes to hcp Mg and Al_2Ca phases for the bulk alloy were obtained by extrusion at $623\,K$. The extruded bulk alloy exhibits rather good mechanical properties of $420–500\,MPa$ for yield strength and $5–8\%$ for elongation [67]. Elongation can be increased significantly without detriment to high strength. The bulk Mg-Ca-Al alloy has a mixed structure of fine hcp Mg grains with a size of about $400\,nm$ including much smaller precipitates with an average size of about $50\,nm$. Yield strengths of $420–500\,MPa$ for the mixed phase alloy with a matrix grain size of about $400\,nm$ agree with the strength which can be expected from the relation between $\sigma_{0.2}$ and $d_g^{-1/2}$ shown in figure 3.21. The Mg-Ca-Al alloy also exhibits a very large elongation of above 700% at $723\,K$ and at a very high strain rate of $0.2\,s^{-1}$. A nearly linear relation exists between flow stress and strain rate. From the linear relation, the strain rate sensitivity exponent (m-value) is found to be 0.45 in the vicinity of $0.2\,s^{-1}$. These data indicate the achievement of high strain rate super-plasticity for the extruded Mg-Ca-Al bulk alloy.

A similar experiment has been made for $Mg_{97}Zn_1Y_2$ and $Mg_{96}Zn_1Y_3$ alloys [9]. Atomized $Mg_{97}Zn_1Y_2$ powder below $32\,\mu m$ in size has a mixed structure consisting mainly of hcp Mg and $Mg_{24}Y_5$ phases. The Mg-Zn-Y alloy powder can be easily consolidated into a bulk form with full density. Figure 3.25 shows the change in density with extrusion temperature for the rapidly solidified Mg-Zn-Y alloys. The density values appear to be almost saturated and no distinct change in density with extrusion temperature is seen. The density is $1.84\,g/cm^3$ for the $2\%\,Y$ alloy and $1.86\,g/cm^3$ for the $3\%\,Y$ alloy. Figure 3.26 shows yield strength and ductility as a function of extrusion temperature for rapidly solidified $Mg_{97}Zn_1Y_2$. As the extrusion temperature rises, the yield strength decreases from $600\,MPa$ to $420\,MPa$ and the ductility increases from 5% to 15%. These mechanical properties are plotted in figure 3.27 [9], including data for rapidly solidified nanocrystalline Mg-Y-Mm, Mg-Zn-Ln, AZ91 and ZK61 alloys and conventional ingot metallurgy Mg-based alloys for comparison [2]. Strengths and ductilities are about 1.5 times higher than those for the former Mg-Y-Mm and Mg-Zn-Ln alloys and about 3 times higher than those for the latter ingot metallurgy alloys. Figure 3.28 shows the specific tensile yield strength of the present

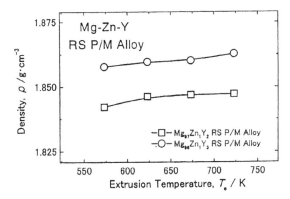

Figure 3.25. Density as a function of extrusion temperature for RS P/M $Mg_{97}Zn_1Y_2$ and $Mg_{96}Zn_1Y_3$ alloys.

Mg-Zn-Y and Mg-Ca-Al alloys in comparison with conventional Mg-, Al- and Ti-based alloys. The specific yield strengths of the new Mg-based alloys are in the range 2.7×10^4–3.3×10^5 Nm/kg which are about 1.2–4 times higher than those for conventional Mg, Al and Ti alloys [68].

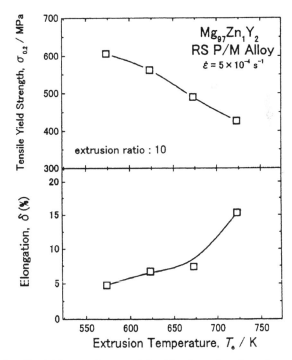

Figure 3.26. Tensile yield strength $\sigma_{0.2}$ and ductility δ as a function of extrusion temperature for the RS P/M $Mg_{97}Zn_1Y_2$ alloy.

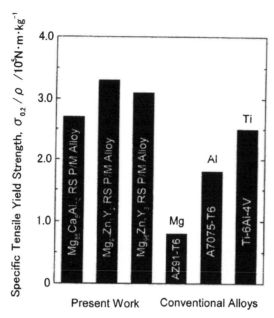

Figure 3.27. Relation between tensile yield strength $\sigma_{0.2}$ and ductility δ for RS P/M Mg-Zn-Y alloys. Data for RS P/M AZ91 and ZK61 alloys, and conventional IM-ageing Mg-based alloys are also shown for comparison.

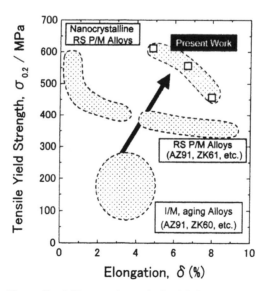

Figure 3.28. Specific tensile yield strength $\sigma_{0.2}/\rho$ for RS P/M Mg-Zn-Y and Mg-Ca-Al alloys. Data for conventional Mg-, Al- and Ti-based alloys are also shown for comparison.

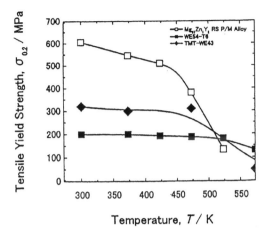

Figure 3.29. Temperature dependence of tensile yield strength $\sigma_{0.2}$ for RS P/M $Mg_{97}Zn_1Y_2$ alloy.

One of the US national projects is to develop Al-based amorphous alloys with higher specific strengths of 300–400 MPa/(Mg/m^3) and higher specific moduli of 20–30 GPa/(Mg/m^3). The specific strength and specific modulus of extruded bulk $Mg_{85}Ca_5Al_{10}$, $Mg_{97}Zn_1Y_2$ and $Mg_{96}Zn_1Y_3$ alloys can be compared with data from conventional Al-, Mg-, Ni- and Ti-based alloys. The newly developed Mg-based alloys have higher specific strength of 270–330 MPa/(Mg/m^3) and high specific modulus of 19–23 GPa/(Mg/m^3). The specific strength values of the new Mg-based alloys are much superior to those for all the conventional metallic materials. Figure 3.29 shows the temperature dependence of the yield strength of the rapidly solidified powder metallurgy $Mg_{97}Zn_1Y_2$ alloy. The Mg-based alloy also exhibits high elevated temperature strength above 400 MPa in the temperature range up to 473 K. Further increase in testing temperature to 523 K causes a significant decrease in yield strength to 130 MPa accompanying a significant increase in elongation to 135%. The yield strength values in the temperature range up to 473 K are much superior to those for the conventional heat resistant Mg-based alloys WE54-T6 and TMT-WE43 [2]. The $Mg_{97}Zn_1Y_2$ alloy also exhibits high strain rate superplasticity as is evidenced from large elongations, above 700% at 623 K in the vicinity of high strain rate of 0.1 s^{-1}, and a high strain rate sensitivity exponent of 0.4. The change in sample shape before and after tensile testing is shown in figure 3.30. The sample with the largest elongation at the high strain rate of 0.10 s^{-1} was elongated through a homogeneous deformation mode. Figure 3.31 shows the relation between elongation and strain rate for the rapidly solidified $Mg_{97}Zn_1Y_2$ alloy, together with data from rapidly solidified AZ91, ZK61 and Mg-Cu-Y alloys [2] as well as 7075-T6 alloy [69]. The ductility is the highest for Mg-Zn-Y and the strain rate leading

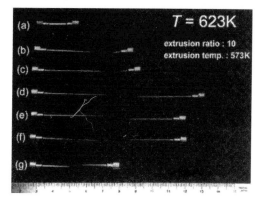

Figure 3.30. Change in the sample morphology of RS P/M $Mg_{97}Zn_1Y_2$ alloy before and after tensile testing at 623 K and various strain rates: (a) before deformation, (b) $2 \times 10^{-2}\,s^{-1}$, (c) $4 \times 10^{-2}\,s^{-1}$, (d) $6 \times 10^{-2}\,s^{-1}$, (e) $1 \times 10^{-1}\,s^{-1}$, (f) $2 \times 10^{-1}\,s^{-1}$, (g) $8 \times 10^{-1}\,s^{-1}$.

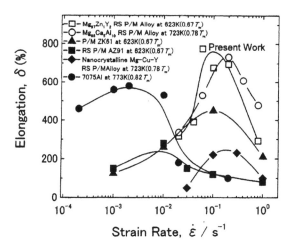

Figure 3.31. Relation between elongation δ and strain rate for RS P/M $Mg_{97}Zn_1Y_2$. Data for other RS P/M Mg- and Al-based alloys are also shown for comparison.

to the highest elongation is also located at the high strain rate side. These results indicate that the Mg-Zn-Y alloy possesses the best high-strain rate superplastic characteristics.

Rapidly solidified Mg-Zn-Y microstructures

Rapidly solidified $Mg_{97}Zn_1Y_2$ alloy has excellent mechanical properties as well as high strain rate superplasticity. The microstructural origin of these

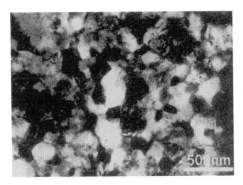

Figure 3.32. Bight-field transmission electron micrograph from RS P/M $Mg_{97}Zn_1Y_2$ alloy obtained by extrusion at 573 K.

good mechanical properties can be examined by transmission electron microscopy [9]. Figure 3.32 shows a transmission electron microscope image of the rapidly solidified $Mg_{97}Zn_1Y_2$ alloy obtained by extrusion at 573 K. The alloy consists of fine grains with a grain size of 100–150 nm. In addition, each grain contains a high density of stacking faults as well as very fine precipitates with a size of about 10 nm. When the specimen is tilted to an optimum Bragg reflection condition, a high density of stacking faults can be seen in all the grains, as exemplified in figure 3.33. This is believed to be the first evidence for the introduction of stacking faults into hcp Mg. In the selected-area electron diffraction pattern shown in figure 3.33, there are extra reflection spots at the positions of $1/3(0001)Mg$ and $2/3(0001)Mg$, indicating the

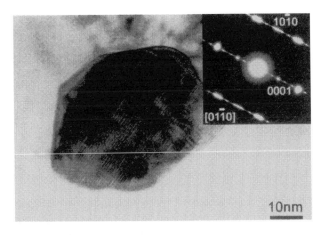

Figure 3.33. Bright-field transmission electron micrograph and selected-area electron diffraction pattern from RS P/M $Mg_{97}Zn_1Y_2$ alloy obtained by extrusion at 573 K.

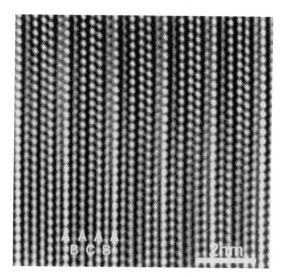

Figure 3.34. High resolution transmission electron micrograph of RS P/M $Mg_{97}Zn_1Y_2$ alloys obtained by extrusion at 573 K.

formation of a long-periodic hexagonal structure in which the periodicity is 3 times longer than the ordinary hcp Mg phase. A high-resolution transmission electron micrograph of the faulted region is shown in figure 3.34. The atomic configuration is ABACAB type, 3 times longer than the ABABAB type [9] which is typical for the ordinary hcp Mg phase. It is therefore concluded that the Mg-Zn-Y solid solution has a novel atomic configuration which is significantly different from hcp Mg. In addition to the change in the structure of the matrix Mg, there is a homogeneous dispersion of fine cubic shape precipitates with an edge size of about 7 nm. These precipitates can be characterized as an $Mg_{24}Y_5$ phase, with a cubic structure of lattice parameter 1.1 nm, on the basis of the high-resolution electron microscope (HREM) image, nanobeam electron diffraction pattern and nanobeam energy dispersive x-ray microanalysis (EDAX) profile shown in figure 3.35.

In the equilibrium phase diagram of the Mg-Zn-Y system [70], there is no appreciable solid solubility limit for Zn and Y in hcp Mg at room temperature. Consequently, the present Mg phase is a reinforced solid solution saturated with Zn and Y. Such a reinforced solid solution generates significant strains because of the atomic size ratios of 14% for Y/Mg and 20% for Mg/Zn [71]. The high density of stacking faults and associated change in atomic configurations is presumed to be introduced to relax these strains. This suggests that simultaneous addition of more than two kinds of elements with significant atomic size ratios may produce a new type of solid solution phase with a novel structure. Overall, the mechanisms for

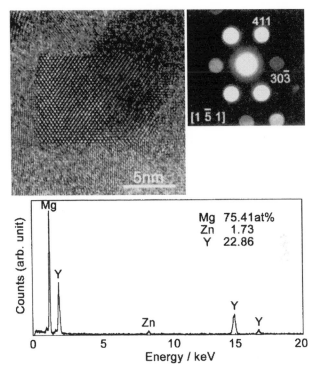

Figure 3.35. High-resolution transmission electron micrograph, nanobeam electron diffraction pattern and nanobeam EDX profile of RS P/M $Mg_{97}Zn_1Y_2$ alloy.

high strength and high ductility in rapidly solidified Mg-Zn-Y alloys [9] result from a combination of the following four factors: (1) nanograin hexagonal structure with grain sizes of 100–150 nm, (2) reinforced solid solution of Y and Zn elements with significantly different atomic sizes, (3) high density of stacking faults in the hexagonal Mg phase leading to the significant change in the atomic configurations, and (4) homogeneous precipitation of cubic $Mg_{24}Y_5$ particles with a size of about 10 nm in the hexagonal Mg matrix.

Conclusions

High-strength Al- and Mg-based alloys have been developed consisting of nanoscale amorphous, quasicrystalline or long periodic hexagonal phases, by controlling the cluster structure, composition and stability of the super-cooled liquid. It is expected that this new synthesis method of controlling the stability, composition and structure of the supercooled liquid can lead to future fabrication of other kinds of Al- and Mg-based materials with high functional properties.

References

[1] Wilm A 1911 *Metallurgie* **8** 225
[2] *Handbook of Advanced Magnesium Technology* 2000 (Tokyo: Kallos Publishing)
[3] Inoue A, Ohtera K, Tsai A P and Masumoto T 1988 *Jpn. J. Appl. Phys.* **27** L280
[4] Kim Y H, Inoue A and Masumoto T 1990 *Mater. Trans. JIM* **31** 747
[5] Inoue A 1994 *Mater. Sci. Eng.* **A179/A180** 57
[6] Inoue A and Kimura H M 1997 *Mater. Sci. Forum* **235–238** 873
[7] Inoue A 1997 *Handbook on the Physics and Chemistry of Rare Earths* vol 24, eds K A Gschneidner and L Eyring p 84
[8] Kato A, Inoue A, Horikiri H and Masumoto T 1995 *Mater. Trans. JIM* **36** 977
[9] Inoue A, Kawamura Y, Matsushita M, Hayashi K and Koike J 2001 *J. Mater. Res.* **16** 1894
[10] He Y, Poon S J and Shiflet G J 1988 *Science* **241** 1640
[11] Tsai A P, Inoue A and Masumoto T 1988 *Metall. Trans.* **19A** 1369
[12] Inoue A, Amiya K, Yoshii I, Kimura H M and Masumoto T 1994 *Mater. Trans. JIM* **35** 485
[13] Inoue A, Onoue K and Masumoto T 1994 *Mater. Trans. JIM* **35** 808
[14] Inoue A, Ohtera K, Tsai A P, Kimura H M and Masumoto T 1988 *Jpn. J. Appl. Phys.* **27** L1579
[15] Inoue A, Kita K, Ohtera K and Masumoto T 1988 *J. Mater. Sci. Lett.* **7** 1287
[16] Inoue A, Kawamura Y and Sasamori K unpublished work
[17] Inoue A 1995 *Mater. Trans. JIM* **36** 866
[18] Inoue A, Zhang T and Takeuchi A 1998 *Mater. Sci. Forum* **269–272** 855
[19] Inoue A, Kimura H M and Kita K 1997 *New Horizons in Quasicrystals* ed. A I Goldman, D J Sordelet, P A Thiel and J M Dubois (Singapore: World Scientific) p 256
[20] Kim Y H, Inoue A and Masumoto T 1991 *Mater. Trans. JIM* **32** 599
[21] Inoue A, Nakazato K, Kawamura Y, Tsai A P and Masumoto T 1994 *Mater. Trans. JIM* **35** 95
[22] Ohtera K, Inoue A, Terabayashi T, Nagahama H and Masumoto T 1992 *Mater. Trans. JIM* **33** 775
[23] *Metals Databook* 1983 ed. Japan Inst. Metals (Tokyo: Maruzen)
[24] Vasudevan A K and Doherty R O 1989 *Aluminum Alloys* (London: Academic Press)
[25] Martin J W 1980 *Micromechanism in Particle Hardened Alloys* (Cambridge: Cambridge University Press) p 40
[26] England R O, Pickens J R, Kumar K S and Langan T J 1988 *Dispersion Strengthened Aluminum Alloys* ed. Y W Kim and W M Griffith (London: Minerals, Metal and Materials Society) p 371
[27] Hall E O 1951 *Proc. Phys. Soc. London* **B64** 747
[28] Petch N J 1953 *J. Iron Steel Inst.* **174** 25
[29] Bennett V R, Nix W D and Tettleman A S 1973 *The Principles of Engineering Materials* (Prentice-Hall) p 226

[30] Decker R F 1973 *Metall. Trans.* **4** 2495
[31] Wert J A 1980 *Strength of Metal Alloys* ed. R C Gifkins (Oxford: Pergamon Press) p 339
[32] Kim Y W and Griffith W M 1984 *PM Aerospace Materials* (Shrewdburg, UK: MPR) p 1
[33] Higashi K, Mukai T, Tanimura S, Inoue A, Masumoto T, Kita K, Ohtera K and Nagahora J 1992 *Scripta Metall.* **26** 191
[34] *YKK Catalog* 1995 Tokyo
[35] Shechtman D, Blech L A, Gratias D and Cahn J W 1984 *Phys. Rev. Lett.* **53** 1951
[36] Inoue A, Kimura H M and Masumoto T 1987 *J. Mater. Sci.* **22** 1758
[37] Rao K V, Fildler J and Chen H S 1986 *Europhys. Lett.* **1** 647
[38] Inoue A, Arnberg L, Lehtinen B, Oguchi M and Masumoto T 1986 *Metall. Trans.* **17A** 1657
[39] Elser Y and Henley C L 1985 *Phys. Rev. Lett.* **55** 2883
[40] Takeuchi S 1992 *Tetsu-to-Hagane* **78** 1517
[41] Voisin E and Pasturel A 1987 *Phil. Mag. Lett.* **55** 123
[42] Yokoyama Y, Tsai A P, Inoue A and Masumoto T 1991 *Mater. Trans. JIM* **32** 1089
[43] Yokoyama Y, Inoue A and Masumoto T 1993 *Mater. Trans. JIM* **34** 135
[44] Inoue A, Watanabe M, Kimura H M, Takahashi F, Nagata A and Masumoto T 1992 *Mater. Trans. JIM* **33** 723
[45] Inoue A, Kimura H M, Sasamori K and Masumoto T 1994 *Mater. Trans. JIM* **35** 85
[46] Inoue A, Kimura H M, Sasamori K and Masumoto T 1995 *Mater. Trans. JIM* **36** 6
[47] Inoue A, Kimura H M, Watanabe M and Kawabata A 1997 *Mater. Trans. JIM* **38** 756
[48] Inoue A and Kimura H M 1997 *Mater. Sci. Forum* **235–238** 873
[49] Inoue A, Kimura H M and Sasamori K 1997 *Chemistry and Physics of Nanostructures and Related Non-Equilibrium Materials* ed. E Ma, B Fultz, R Shull, J Morral and P Nash (London: The Minerals, Metals and Materials Society) p 201
[50] Inoue A, Kimura H M, Sasamori K and Kita K 1998 *Aluminum Alloys* **3** 1841
[51] Inoue A, Kimura H M and Sasamori K 1998 *Adv. Mater.* **4** 91
[52] Vasudevan A K and Doherty R O 1989 *Aluminum Alloys* (London: Academic Press)
[53] Inoue A, Kimura H M and Masumoto T 1996 *Nanostruct. Mater.* **7** 363
[54] Kimura H M, Sasamori K and Inoue A 1996 *Mater. Trans. JIM* **37** 1722
[55] St. Amand R and Giessen B C 1978 *Scripta Metall.* **12** 1021
[56] Calka A and Matyja 1978 *Proc. Amorphous Metallic Metals* p 71
[57] Inoue A, Ohtera K, Kita K and Masumoto T 1988 *Jpn. J. Appl. Phys.* **27** L2248
[58] Kim S G, Inoue A and Masumoto T 1990 *Mater. Trans. JIM* **31** 929
[59] Inoue A, Nakamura T, Nishiyama N and Masumoto T 1992 *Mater. Trans. JIM* **33** 937
[60] Amiya K and Inoue A 2001 *Mater. Trans. JIM* **2** 543
[61] Kim S G and Inoue A 1991 unpublished research
[62] Kato A 1995 Doctoral Thesis, Tohoku University
[63] Fisher J C, Hart E W and Pry R H 1953 *Acta Metall.* **1** 336
[64] Inoue A 1998 *Bulk Amorphous Alloys* (Zurich: TransTech Publications) p 1
[65] Inoue A 2000 *Acta Mater.* **48** 279
[66] Hayashi K, Kawamura Y and Inoue A unpublished work
[67] Hayashi K, Kawamura Y and Inoue A unpublished work
[68] *Metal Databook* 1983 ed. Japan Inst. Metals (Tokyo: Maruzen)

[69] Kimura H M, Inoue A, Sasamori K and Kawamura Y 1997 J. Jpn. Inst. Light Met. **47** 487

[70] Villars P, Prince A, Zabdyr L and Moser Z 1995 *Handbook of Ternary Alloy Phase Diagrams* (Metals Park, OH: ASM International) p 12369

[71] *Metals Databook* 1983 ed. Japan Inst. Metals (Tokyo: Maruzen) p 8

Chapter 4

Nanocrystallization in Al alloys

Dmitri Louzguine and Akihisa Inoue

Introduction

This chapter summarizes data obtained so far on the influence of a supercooled liquid region on crystallization behaviour in Al-RE-Ni-Co (RE = rare-earth metal) metallic glasses produced by rapid solidification of the melt. It also describes the crystallization process of the Al-based metallic glasses for different regimes of heat treatment. The crystallization behaviour of the Al-based metallic glasses follows different transformation mechanisms above and below the glass-transition temperature. The formation of primary nanoscale α-Al particles is observed during continuous heating and after annealing above the glass-transition temperature. During isothermal annealing below the glass-transition temperature an unknown metastable phase is formed primarily or conjointly with α-Al. $Al_{85}RE_8Ni_5Co_2$ amorphous alloys exhibit no glass-transition and crystallize similarly for all heat treatment regimes, i.e. during isothermal calorimetry at different temperatures and during continuous heating using differential scanning calorimetry at different heating rates. The supercooled liquid range in the $Al_{85}RE_8Ni_5Co_2$ alloys strongly depends upon the electronegativity of the rare earth metal component.

Al-based metallic glasses and nanomaterials

Devitrification of metallic glasses is a powerful tool for creating nanomaterials. A large variety of metallic glasses, including a large number of Al-based ones, undergo nanocrystallization, i.e. the precipitation of nanoscale particles from the glassy phase. Nanocrystallization mostly takes place by either primary crystallization or spinodal decomposition. Nanocrystallization by primary crystallization needs a high nucleation rate to form a large number of crystals, and long-range diffusion to redistribute the solute elements and inhibit crystal growth. The resulting nanomaterials contain nanoscale crystalline (or quasicrystalline at certain compositions) grains or particles separated by high-angle boundaries or homogeneously dispersed in the residual amorphous matrix.

A wide variety of Al-based metallic glasses has been developed to date [1, 2]. Some of these alloys exhibit remarkable mechanical properties, for example high tensile fracture strength exceeding 1000 MPa, which is unattainable in conventional aluminium alloys. Al-based metallic glasses also possess high specific strength due to their low density, and are expected therefore to be used as high-strength structural materials, mainly for aeroplane and aerospace applications. Ternary Al-RE-TM alloys (RE = rare earth metal and TM = transition metal), were discovered in the late 1980s in Japan [3, 4] and the USA [5], and form one of the most important groups of Al-based metallic glasses discovered to date [6]. Simultaneous alloying with a rare earth and transition metal causes drastic widening of the composition range of the amorphous single phase [3, 4]. The alloys are usually produced in a ribbon shape by the melt spinning technique [7] or in a powder form by gas atomization [8]. It is important to note that ternary Al-RE-TM amorphous alloys, e.g. Al-Y-Ni, in addition to high strength also possess good bending ductility (i.e. have the ability to be bent through 180° without fracture) [9, 10]. Addition of Co, partially replacing Y in $Al_{85}Y_{10}Ni_5$ (all alloy compositions are given in nominal at%), increases the tensile fracture strength from 920 MPa to 1250 MPa, in the case of $Al_{85}Y_8Ni_5Co_2$ metallic glass, without worsening the bending ductility [11].

However, many of these Al based glasses can only be manufactured up to 1 mm thick, because of their ease of crystallization, high density of quenched-in nuclei, and low reduced glass-transition and crystallization temperatures [12]. However, addition of 2 at% Co to $Al_{85}Y_{10}Ni_5$ alloy causes a drastic increase in glass-forming ability. The critical ribbon thickness increases from 120 to 900 μm [11]. Moreover, $Al_{85}Ni_{10}Ce_5$ bulk amorphous samples of high relative density are obtained by warm extrusion of atomized amorphous powder [13]. At the same time, the high-strength quaternary $Al_{85}Y_8Ni_5Co_2$ alloy [11] shows one of the widest supercooled liquid region on heating among Al-based metallic glasses.

Different Al-RE-TM glasses containing about 85 at% Al show primary precipitation of fcc α-Al solid solution nanoparticles on heating [14, 15], with extremely high nucleation rates, above 10^{20} m^{-3} s^{-1} [16, 17]. Measurements of the α-Al lattice parameter and atom probe ion field microscopy investigations [18] show very low concentrations of the alloying elements in the crystalline α-Al, in accordance with Al-RE and Al-TM phase diagrams [19]. Segregation of low diffusivity rare-earth metals on the α-Al/amorphous interface [18] is considered to be one of the most important reasons for the low growth rate of the α-Al crystals. Overall, a high number density of α-Al grains is formed by heterogeneous nucleation on a high density of quenched-in nuclei formed during rapid solidification [20], and a primary crystallization process requires long-range diffusion and impedes crystal growth [21, 22].

Deformation-induced nanocrystallization has also been observed in an Al-Fe-Gd metallic glass [23]. Moreover, Cu in $Al_{88}Y_4Ni_7Cu_1$ [24] and $Al_{85}Y_8Ni_{5-4}Cu_{1-2}$ [25] alloys induces the formation of primary α-Al nano-crystals directly from the melt on rapid solidification.

The crystallization behaviour of various Al-Y-Ni-Co alloys has been studied, and at definite compositions formation of α-Al was observed at the primary stage [11, 26, 27]. Addition of a fifth element, for example Zr, makes the crystallization process more complicated but α-Al is still the primary crystalline phase [28].

As well as in the case of the ternary Al-Y-Ni alloys, fine precipitates of α-Al formed by devitrification of a $Al_{85}Y_8Ni_5Co_2$ glassy matrix strengthen the alloy without reduction of bending ductility [29]. An optimum strength value is obtained for a crystalline phase volume fraction ranging from 5 to 25% [30]. Two theories have been put forward to describe this phenomenon. The first theory suggests that the increase in tensile strength by precipitation of α-Al nanoparticles in the amorphous matrix results from a combination of the following three effects:

1. a defect-free effect, with the α-Al particles free of linear defects;
2. an interface effect, with the amorphous/Al particle interface having highly dense packed atomic configurations without excess vacancies and a low interfacial energy; and
3. a nanoscale effect, with the α-Al particle size smaller than the width of the inhomogeneous shear bands, so they can act as effective barriers against the shear deformation of the amorphous matrix [30].

This theory also explains the retention of good bending ductility. A second theory has also been suggested, with hardening attributed mainly to solute enrichment of the residual glassy matrix due to lowering of the Al content [31].

Thus, Al-RE-TM metallic glasses and nanomaterials form one of the most attractive groups of the Al-based nonequilibrium materials. Strengthening of metallic glasses by α-Al nanoparticles means that devitrification studies are important in order to determine optimal regimes of heat treatment from the commercial viewpoint. The rest of this chapter concentrates on recent studies of full or partial replacement of Y in $Al_{85}Y_8Ni_5Co_2$ by other RE metals, namely La, Sm and Mischmetal (Mm), and their effect on the supercooled liquid and nanocrystallization [32–34].

Influence of electronegativity of RE metal on the supercooled liquid region

$Al_{85}RE_8Ni_5Co_2$ and $Al_{85}Y_{8-x}RE_xNi_5Co_2$ alloys (RE excludes Y) form

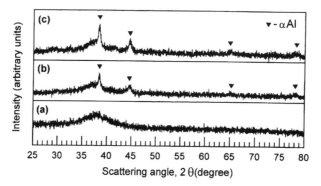

Figure 4.1. XRD patterns (Cu K_α) from (a) $Al_{85}Dy_8Ni_5Co_2$ alloy in as-solidified state, (b, c) alloys heat treated at the temperature of the first exothermic peak: (b) $Al_{85}Gd_8Ni_5Co_2$ alloy at 581 K for 0.9 ks, (c) $Al_{85}Dy_8Ni_5Co_2$ alloy at 581 K for 0.9 ks.

metallic glasses upon rapid solidification. For example, an x-ray diffraction pattern (XRD) of the as-solidified $Al_{85}Dy_8Ni_5Co_2$ alloy is shown in figure 4.1(a). Differential scanning calorimetry (DSC) traces of $Al_{85}RE_8Ni_5Co_2$ glasses shown in figure 4.2 reveal glass transition and a number of exothermic peaks on heating. The onset temperatures of glass-transition (T_g) and crystallization (T_x) shown in figure 4.2 can be determined by the intersection point of two tangents to the curve below and above the onset temperature. It is seen that Mm- and La-bearing alloys exhibit no glass transition. The Nd-bearing alloy shows the very beginning of glass-transition, but it is impossible to estimate the $\Delta T_x = (T_x - T_g)$ temperature interval. Other RE-bearing alloys (see figure 4.2) exhibit clear glass-transitions.

The heat of mixing for binary liquid systems at equiatomic compositions, an important factor influencing glass-forming ability and stability of the supercooled liquid [32], is calculated to be -38 kJ/mol [33, 35] for all Al-RE (including Y) atomic pairs. However, a great difference in the influence of various RE metal additions on ΔT_x is observed in the $Al_{85}RE_8Ni_5Co_2$ alloys as shown in table 4.1. From this point it is not clear why addition of Y having no f-electrons (see table 4.1) makes the same influence on ΔT_x as other RE metals with a large number of f-electrons. Table 4.1 also contains Pauling electronegativity data [34] for the RE metals.

Figure 4.3 demonstrates that ΔT_x increases almost linearly with the electronegativity of the constituent RE metal [35]. The ΔT_x increases corresponds to a reduction in T_g and a slight increase in T_x. Alloying with Ce-based Mischmetal [39] as well as with La causes the supercooled liquid region to disappear, because of a low Ce electronegativity of 1.12. The data point corresponding to the Nd-bearing alloy is not shown due to its uncertainty. The difference in electronegativity of the constituent elements is an important factor influencing glass formation because it affects the

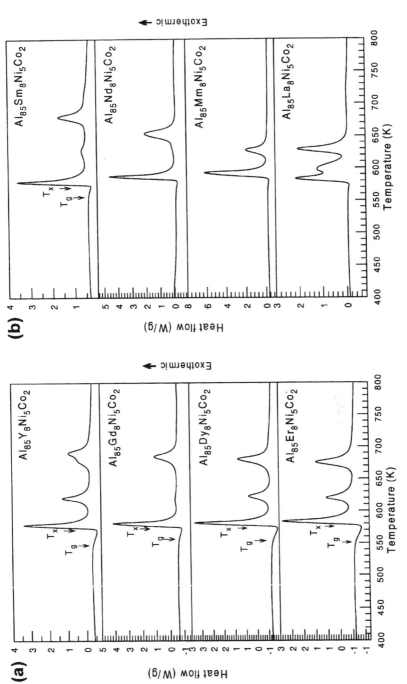

Figure 4.2. DSC traces taken at 0.67 K/s.

Nanocrystallization in Al alloys

Table 4.1. Electronic configurations and Pauling electronegativities of the RE metal and the range of the supercooled liquid region (ΔT_x) in $Al_{85}RE_8Ni_5Co_2$ alloys.

Element	Electronic configuration	Pauling electronegativity (Pauling units)	ΔT_x (K)
^{39}Y	$[Kr]\,4d^1 5s^2$	1.22	29
^{57}La	$[Xe]\,5d^1 6s^2$	1.10	—
^{60}Nd	$[Xe]\,4f^4 6s^2$	1.14	*
^{62}Sm	$[Xe]\,4f^6 6s^2$	1.17	15
^{64}Gd	$[Xe]\,4f^7 5d^1 6s^2$	1.20	21
^{66}Dy	$[Xe]\,4f^{10} 6s^2$	1.22	26
^{68}Er	$[Xe]\,4f^{12} 6s^2$	1.24	31

* Almost zero, difficult to estimate.

interatomic constraint forces. Electronegativity is the power of a bound atom to attract electrons.

Al is amphoteric with a location close to the metalloids in the periodic table. This suggests partial localization of the electrons in Al-RE-Ni-Co

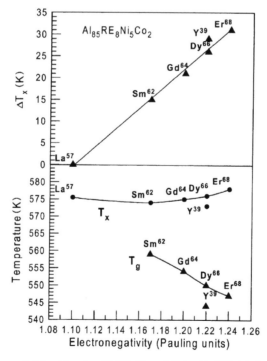

Figure 4.3. Supercooled liquid region (ΔT_x), T_g and T_x in $Al_{85}RE_8Ni_5Co_2$ metallic glasses as functions of the electronegativity of the RE metal.

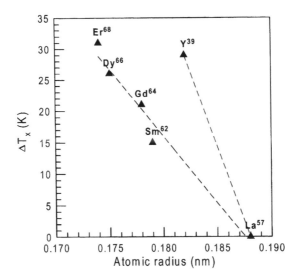

Figure 4.4. ΔT_x in $Al_{85}RE_8Ni_5Co_2$ metallic glasses as a function of the atomic radius of the RE metal.

alloys, leading to the formation of a partially ionic bond between the alloying elements that influences formation of the supercooled liquid on heating. This indicates the importance of studying the electronic structure in Al-RE-Ni-Co alloys, e.g. by computer simulation, which is however hampered due to the complicated electronic structure of RE metals. Partial replacement of Y by Zr with significantly higher electronegativity (1.33), closer to that of Al (1.61), causes primary crystallization of I4/*mmm* Al_3Zr during rapid solidification [33]. This is despite the slightly higher heat of mixing for the Al-Zr atomic pair ($-44\,kJ/mol$) than $-38\,kJ/mol$ [37] obtained for the Al-Y and some other Al-RE atomic pairs. This suggests that there is an optimal value of electronegativity of the alloying element in the $Al_{85}RE_8Ni_5Co_2$ alloys that leads to stabilization of the supercooled liquid. Formation of the supercooled liquid seems to be connected with the electronic structure of the alloy. Another important factor for glass formation is the difference in atomic radii between the alloying elements, which is determined by the atomic radius of the RE metal, but does not correlate well with ΔT_x (see figure 4.4). For example, Nd has the same atomic size as Y but a great difference in ΔT_x, as shown in figure 4.2.

Crystallization above and below the glass transition temperature

Different Al-based metallic glasses exhibit formation of primary α-Al nanoparticles upon crystallization. Y-, Sm-, Gd- and Dy- bearing metallic glasses

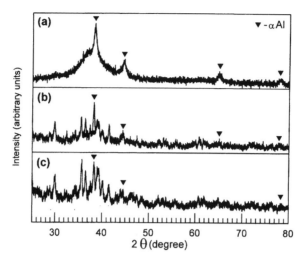

Figure 4.5. XRD patterns from $Al_{85}Ni_5Y_4Nd_4Co_2$ after isothermal calorimetry (a) at 568 K for 0.36 ks, (b) at 543 K for 2 ks and (c) at 533 K for 5.4 ks.

with the $Al_{85}RE_8Ni_5Co_2$ composition show precipitation of α-Al nanoparticles after continuous heating using differential scanning calorimetry at high enough heating rate (0.67 K/s and higher) or annealing at temperatures above T_g, as shown in figures 4.1(b) and (c). For these alloys a heating rate of 0.67 K/s is high enough to prevent crystallization below T_g. An identical behaviour is found in the $Al_{85}Y_4Nd_4Ni_5Co_2$ metallic glass, as shown in figure 4.5(a). However, Y- [36], Gd- and Dy-bearing metallic glasses as well as $Al_{85}Y_4Nd_4Ni_5Co_2$ [34] show simultaneous formation of intermetallic compound(s) and α-Al nanoparticles or primary formation of an intermetallic compound after annealing below T_g.

For example, the $Al_{85}Y_8Ni_5Co_2$ alloy shows formation of an unknown intermetallic compound conjointly with α-Al nanoparticles after annealing up to the completion of the primary phase transformation, as shown in figures 4.6 and 4.7. Several reflections belonging to the intermetallic compound phase coincide with the (1 1 1) ring of α-Al, indicating possible formation of a semicoherent interface between the intermetallic compound and α-Al. After completion of the primary phase transformation, many relatively small intermetallic compound particles below 50 nm size exist in the structure of the $Al_{85}Y_8Ni_5Co_2$ alloy, as shown in figure 4.7. However, some of the particles reach a size of about 0.5 μm, as shown in figure 4.8, especially at higher annealing temperature but not exceeding T_g. Another intermetallic compound precipitates below T_g in the $Al_{85}Y_4Nd_4Ni_5Co_2$ alloy [34]. Both intermetallic compounds are metastable and have multicomponent compositions. The volume fraction of the intermetallic compound is higher than that of α-Al and the fraction of α-Al depends upon an annealing temperature below T_g,

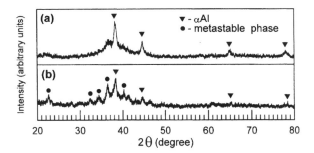

Figure 4.6. XRD patterns from $Al_{85}Y_8Ni_5Co_2$ after (a) DSC at 0.67 K/s up to the completion of the first exothermic reaction, i.e. up to 590 K, and (b) isothermal calorimetry below T_g up to the completion of the first exothermic reaction, i.e. for 5.4 ks at 533 K.

as shown in figures 4.5(b) and (c). The multicomponent intermetallic phase is the leading phase during isothermal crystallization of $Al_{85}Y_4Nd_4Ni_5Co_2$ below T_g, i.e. it starts to precipitate prior to α-Al.

The data shown in figures 4.5-4.8 illustrate differences in crystallization mechanism and products below and above T_g. Johnson–Mehl [37] and Avrami [38] analysis as well as that done by Kolmogorov [39] is used to study crystallization kinetics when phase transformation takes place by nucleation and growth. The fraction transformed x is given as a function of time t by

$$x(t) = 1 - \exp[-Kt^n] \qquad (4.1)$$

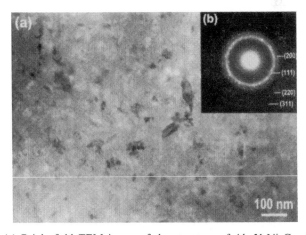

Figure 4.7. (a) Bright-field TEM image of the structure of $Al_{85}Y_8Ni_5Co_2$ heat treated using isothermal calorimetry at 533 K for 5.4 ks, and (b) selected-area electron diffraction pattern, with sharp rings from α-Al as indexed and other spots from an intermetallic phase. JEM 2010 microscope operating at 200 kV [40].

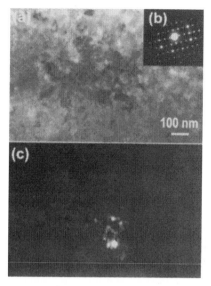

Figure 4.8. The structure of $Al_{85}Y_8Ni_5Co_2$ heat treated using isothermal calorimetry at 533 K for 5.4 ks: (a) bright-field image, (b) selected-area electron diffraction pattern and (c) dark-field image.

which describes a phase transformation where the nucleation and growth rates are independent of time. Equation (4.1) can also be written as

$$\ln[\ln(1/(1 - x)] = \ln(K) + n\ln(t). \qquad (4.2)$$

Avrami plots of $\ln[-\ln(1 - x)]$ versus $\ln(t)$ can be constructed using isothermal calorimetry data [40], as shown in figure 4.9. During isothermal calorimetry the samples are heated up to testing temperature at the highest possible heating rate of 1.67 K/s. From the differential scanning calorimetry trace taken at 1.67 K/s the glass-transition temperature is found to be 561 K.

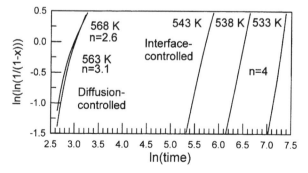

Figure 4.9. Avrami plot for $Al_{85}Ni_5Y_4Nd_4Co_2$ [44].

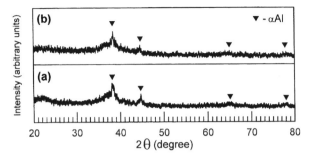

Figure 4.10. XRD patterns of $Al_{85}Ni_5Y_4Sm_4Co_2$: (a) continuous heating using DSC taken at 0.67 K/s up to completion of the first heat effect (i.e. up to 590 K) and (b) isothermally at 543 K for 1 ks [44].

Below T_g the Avrami plot is nearly linear and at 538 K it gives an Avrami exponent of $n = 4$, corresponding to three-dimensional interface-controlled growth of the unknown intermetallic compound and the α-Al phase [45]. As the isothermal calorimetry temperature exceeds 561 K, a change from interface-controlled to diffusion-controlled growth of α-Al nanoparticles takes place, as shown in figures 4.5(b) and (c). The average Avrami exponent n decreases to $n = 2.5$, as shown in figure 4.9, corresponding to three-dimensional diffusion controlled growth of α-Al nanoparticles [41]. Non-linearity of the Avrami plot above T_g illustrates non-steady state nucleation with a time dependent nucleation rate [42].

Exceptional cases are $Al_{85}Y_4Sm_4Ni_5Co_2$ and $Al_{85}Er_8Ni_5Co_2$ alloys. $Al_{85}Y_4Sm_4Ni_5Co_2$ shows formation of only primary α-Al nanoparticles even with isothermal annealing below T_g, as shown in figure 4.10. $Al_{85}Er_8Ni_5Co_2$ shows formation of the intermetallic compound conjointly with α-Al after annealing above T_g as well as after heating using differential scanning calorimetry at 0.67 K/s up to the completion of the first phase transformation related to the primary exothermic peak (see figure 4.2). The Avrami plot for the $Al_{85}Y_4Sm_4Ni_5Co_2$ alloy gives an average n value of 2.7, as shown in figure 4.11, that corresponds to transient nucleation and three-dimensional diffusion-controlled growth of α-Al nanoparticles. This suggests that Sm addition destabilizes the intermetallic compound.

Alloys with no supercooled liquid region, e.g. those with La, Mm and Nd, show similar crystallization products when annealed isothermally about 30–40 K below crystallization temperature or with continuous heating, i.e. primary intermetallic compounds or intermetallic compounds plus α-Al [33] as shown in figure 4.12 [43]. This suggests an important role of the supercooled liquid in crystallization of the Al-based glasses. For various metallic glasses having a wide supercooled liquid region, the viscosity, density, diffusion coefficients and other properties of the super-cooled liquid are different from the as-solidified glassy phase [44]. The

Nanocrystallization in Al alloys

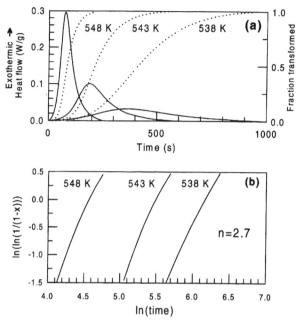

Figure 4.11. Isothermal calorimetry data from $Al_{85}Y_4Sm_4Ni_5Co_2$: (a) solid lines: isothermal calorimetry traces taken at three different temperatures (without incubation period); corresponding dotted lines: fraction transformed as a function of time obtained from the isothermal calorimetry trace. (b) Avrami plot [44].

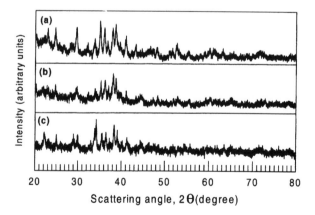

Figure 4.12. XRD patterns from $Al_{85}Nd_8Ni_5Co_2$: (a) after annealing at 590 K for 20 min, (b) after DSC up to the completion of the first exothermic reaction taken at 0.67 K/s and (c) after completion of the primary exothermic reaction using isothermal calorimetry at 552 K [44].

supercooled liquid has a higher degree of dense randomly packed atomic configurations and high solid/liquid interfacial energy, which are favourable for the suppression of nucleation of crystalline phases. The Al-bearing intermetallic compounds have low nucleation and growth rates in the supercooled liquid region, so that only primary α-Al particles precipitate by diffusion-controlled growth. Such behaviour is different from TM-based metallic glasses where diffusion-controlled growth is found below T_g and nucleation and growth above T_g can be described by the viscous flow [45]. The crystallization products of Ti-, [46, 47] and Hf-based [48] and many other metallic glasses are independent of T_g.

Transformation behaviour similar to that observed below T_g in Al-based metallic glasses is found in the Gα-Al-Cr-Ce-Sm amorphous alloy where a multicomponent metastable phase forms conjointly with Fd3m Ge by the initial crystallization reaction [49]. Another similarity is that phase transformation in $Ge_{68}Cr_{14}Al_{10}Ce_4Sm_4$, $Al_{85}Y_4Nd_4Ni_5Co_2$ and $Al_{85}Y_8Ni_5Co_2$ alloys is not complete at the first stage, i.e. a residual amorphous phase remains in the structure, and there is a temperature gap between primary and subsequent phase transformations, as shown by the separation of the first and subsequent exothermic peaks in figure 4.2. However, the Gα-Al-Cr-Ce-Sm amorphous alloy shows a single-stage transformation obeying the kinetic equation (4.1) with $n = 2.8$ [53]. The phase transformation scheme

$$\text{metallic glass} \longrightarrow \text{intermetallic phase} + \alpha\text{-Al}$$

is similar to an irregular eutectic type mechanism (see figures 4.7 and 4.8). However, a common interface between the two product phases is required for coupled eutectic growth. The α-Al and intermetallic phases nucleate separately and do not often have a common interface. Formation of the intermetallic phase enriches the amorphous matrix in Al, which stimulates precipitation of α-Al.

Differences in crystallization mechanism have also been observed in an $Al_{92}Sm_8$ amorphous alloy prepared by rapid solidification of the melt and by solid-state processing (cold rolling) at ambient temperature. No α-Al nanocrystals are formed in material produced by cold rolling, while a large number density of α-Al particles is obtained by devitrification of melt-spun material [50]. Crystallization products in Al-Ni-Y metallic glasses are also found to depend on cooling rate during melt spinning [17]. As with $Al_{85}Y_4Nd_4Ni_5Co_2$, high average Avrami exponent values ranging from 4.3 to 5.4 are obtained for $Al_{85}Dy_8Ni_5Co_2$ [47], indicating interface-controlled conjoint growth of stable R3m (hR20) Al_3Dy, AlDyNi and α-Al phases with increasing nucleation rate. Avrami plots are not linear and non-steady state nucleation is assumed. Isothermal crystallization of the $Al_{85}Dy_8Ni_5Co_2$ metallic glass differs from that of the $Al_{85}Y_8Ni_5Co_2$ and $Al_{85}Y_4Nd_4Ni_5Co_2$ glasses, where formation of metastable intermetallic phases is found.

From an Arrhenius plot of incubation time, the activation energy for nucleation of the $Al_{85}Dy_8Ni_5Co_2$ metallic glass is found to be 360 kJ/mol [44]. High activation energy values of 400 and 300 kJ/mol for $Al_{85}Ni_5Y_4Nd_4Co_2$ and $Al_{85}Y_8Ni_5Co_2$ alloys also indicate a high barrier for nucleation of intermetallic compounds below T_g. $Al_{85}Nd_8Ni_5Co_2$ alloy has no glass-transition and has a lower activation barrier for nucleation of 270 kJ/mol.

Summary

$Al_{85}RE_8Ni_5Co_2$ metallic glasses are prospective materials for different applications as structural materials due to their high strength and relatively low density. Their properties can be improved and tailored by primary α-Al nanocrystallization. The supercooled liquid influences crystallization of $Al_{85}RE_8Ni_5Co_2$ and $Al_{85}Y_{8-x}RE_xNi_5Co_2$ metallic glasses.

Crystallization of most $Al_{85}RE_8Ni_5Co_2$ metallic glasses takes place:

1. by precipitation and diffusion controlled growth of nanoscale α-Al particles during isothermal annealing at temperatures above T_g or during continuous heating in a differential scanning calorimeter; and
2. by interface-controlled growth of primary intermetallic compounds or joint growth of intermetallic compounds and α-Al during isothermal annealing below T_g.

However, $Al_{85}RE_8Ni_5Co_2$ amorphous alloys exhibiting no glass-transition crystallize similarly with different heat treatments, forming primary intermetallic compounds or joint intermetallic compounds and α-Al. The change from interface-controlled to diffusion-controlled growth is observed directly on isothermal annealing of the $Al_{85}Y_4Nd_4Ni_5Co_2$ metallic glass. This behaviour is presumed be observed in both Al-RE-Ni-Co and other Al-RE-TM glasses.

The supercooled liquid region in $Al_{85}RE_8Ni_5Co_2$ alloys depends strongly on the electronegativity of the RE metal. There is an optimal value of electronegativity in $Al_{85}RE_8Ni_5Co_2$ alloys that leads to stabilization of the supercooled liquid. This shows the importance of electronic structure studies of Al-RE-Ni-Co alloys.

References

[1] Inoue A 1997 in *Handbook on the Physics and Chemistry of Rare Earths* vol 24 ed. K A Gschneidner Jr and L Eyring (Amsterdam: North-Holland) p 83
[2] Inoue A 1998 *Prog. Mater. Sci.* **43** 365
[3] Inoue A, Ohtera K, Tsai A P and Masumoto T 1988 *Jpn. J. Appl. Phys.* **27** L280

[4] Inoue A, Ohtera K and Masumoto T 1988 *Jpn. J. Appl. Phys.* **27** L1796

[5] He Y, Poon S J and Shiflet G J 1988 *Science* **241** 1640

[6] Inoue A, Zhang T, Kita K and Masumoto T 1989 *Mater. Trans. JIM* **30** 870

[7] Elliot S R 1990 *Physics of Amorphous Materials* (Harlow: Longman Group) p 139

[8] Inoue A and Kimura H M 2001 *J. Metastable Nanocryst. Mater.* **9** 41

[9] Inoue A, Ohtera K, Tsai A P and Masumoto T 1988 *Jpn. J. Appl. Phys.* **27** L479

[10] Shihlet G J, He Y and Poon S J 1988 *J. Appl. Phys.* **64** 6863

[11] Inoue A, Matsumoto N and Masumoto T 1990 *Mater. Trans. JIM* **31** 493

[12] Allen D R, Foley J C and Perepezko J H 1998 *Acta Mater.* 46 431

[13] Inoue A and Kimura H M 2001 *J. Light Met.* **1** 31

[14] Kim Y H, Inoue A and Masumoto T 1991 *Mater. Trans. JIM* **32** 331

[15] Gogebakan M, Warren P J and Cantor B 1997 *Mater. Sci. Eng. A* **226–228** 168

[16] Foley J C, Allen D R and Perepezko J H 1996 *Scripta Mater.* **35** 655

[17] Calin M, Rudiger A and Koester U 2000 *J. Metastable Nanocryst. Mater.* **8** 359

[18] Hono K, Zhang Y, Tsai A P, Inoue A and Sakurai T 1995 *Scripta Mater.* **32** 191

[19] Massalski T B 1990 *Binary Alloy Phase Diagrams* (Materials Park, OH: ASM International) p 182

[20] Wu R I, Wilde G and Perepezko J H 2001 *Mater. Sci. Eng. A* **301** 12

[21] Yavari A R and Negri D 1997 *Nanostr. Mater.* **8** 969

[22] Perepezko J H, Hebert R J and Tong W S 2002 *Intermetallics* **10** 1079

[23] Jiang W H, Pinkerton F E and Atzmon M 2003 *Scripta Mater.* **48** 1195

[25] Hong S J, Warren P J and Chun B S 2001 *Mater. Sci. Eng. A* **304–306** 362

[25] Louzguine D V and Inoue A 2002 *J. Mater. Res.* **17** 1014

[26] Pekala K, Jaskiewicz P, Latuch J and Kokoszkiewicz A 1997 *J. Non-Cryst. Solids* **211** 72

[27] Bassim N, Kiminami C S and Kaufman M J 2000 *J. Non-Cryst. Solids* **273** 271

[28] Sá Lisboa R D and Kiminami C S 2002 *J. Non-Cryst. Solids* **304** 36

[29] Kim Y H, Inoue A and Masumoto T 1991 *Mater. Trans. JIM* **32** 599

[30] Inoue A, Kimura H M and Kita K 1997 *New Horizons in Quasicrystals* ed. A I Goldman, D J Sordelet, P A Thiel and J M Dubois (Singapore: World Scientific) p 256

[31] Greer A L 2001 *Mater. Sci. Eng. A* **304–306** 68

[32] Inoue A 2000 *Acta Mater.* **48** 279

[33] Takeuchi A and Inoue A 2000 *Mater. Trans. JIM* **41** 1372

[34] James A M and Lord M P 1992 in *Macmillan's Chemical and Physical Data* (London: Macmillan) p 1020

[35] Louzguine D V and Inoue A 2001 *Appl. Phys. Lett.* **79** 3410

[36] Louzguine D V and Inoue A 2002 *Mater. Lett.* **54** 75

[37] Johnson M W A and Mehl K F 1939 *Trans. Am. Inst. Mining Met. Eng.* **135** 416

[38] Avrami M 1941 *J. Chem. Phys.* **9** 177

[39] Kolmogorov A N 1937 *Isv. Akad. Nauk. USSR, Ser. Matem.* **3** 355 [in Russian]

[40] Louzguine D V and Inoue A 2002 *J. Non-Cryst. Solids* **311** 281

[41] Christian J W 1975 *The Theory of Transformations in Metals and Alloys* (Oxford: Pergamon Press) p 542

[42] Kelton K F 2000 *Acta Mater.* **48** 1967

[43] Louzguine D V and Inoue A 2002 *Proc. ISMANAM 2001 Int. Symp. on Metastable, Mechanically Alloyed and Nanocrystalline Materials, Ann Arbor, Michigan, USA, 24–29 June 2001; Mater. Sci. Forum* **386–388** 117

[44] Inoue A 2000 *Acta Mater.* 48 279
[45] Köster U and Meinhardt J 1994 *Mater. Sci. Eng. A* **178** 271
[46] Louzguine D V and Inoue A 2000 *J. Mater. Sci.* **35** 4159
[47] Louzguine D V and Inoue A 2000 *Scripta Mater.* **43** 371
[48] Louzguine D V, Ko M S and Inoue A 2001 *Scripta Mater.* **44** 637
[49] Louzguine D V and Inoue A 2000 *J. Mater. Sci.* **35** 5537
[50] Wilde G, Sieber H, Perepezko J H 1999 *J. Non-Cryst. Solids* **250** 621

Chapter 5

High strength nanostructured Al-Fe alloys

Kazuhiko Kita, Hiroyuki Sasaki, Junichi Nagahora
and Akihisa Inoue

Introduction

Recently, improvements in the strength of lightweight metals have become important because of demands to save energy and increase energy efficiency. The strengthening of aluminium alloys is important, since they are the second most widely used metals after steels. Refinement of grain size is one of the most powerful strengthening mechanisms, which can be achieved by increasing the rate of solidification. At the same time, it is possible to extend solid solution limits for alloying elements, thus providing further increases in strength. Cooling rates in conventional casting, however, are low, being typically in the range 10–100 K/s [1]. Compared with casting, physical vapour deposition (PVD), which is commonly used for fabrication of thin films and coatings, and liquid quenching techniques have much higher cooling rates [1]. Liquid-quenched Al-based alloys with an amorphous phase or nanocrystalline microstructure are reported to have high strengths of 1000–1500 MPa [2]. The forms of products from the liquid-quenching process are powders or ribbons, which require hot working during consolidation to form a bulk product. During consolidation, however, crystallization and grain coarsening lead to a deterioration of the mechanical properties in comparison with those in the as-quenched state. On the other hand, thin films produced by physical vapour deposition often have nonequilibrium phase structures [3]. Physical vapour deposition enables bulk production of alloys directly without consolidation. This chapter discusses the microstructure and mechanical properties of bulk Al-Fe deposited alloys prepared by dual source electron beam evaporation.

   Al-based alloys with additions of Fe have been prepared by physical vapour deposition using dual electron beam evaporation sources. The electron-beam evaporation process has the highest rate of deposition compared with other physical vapour deposition techniques. A continuous

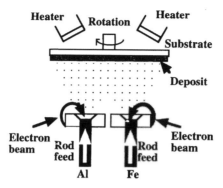

Figure 5.1. Schematic diagram of continuous electron-beam vacuum chamber setup.

supply of rod-shaped evaporation sources produces a bulky deposit of more than 1.5 mm in thickness. A schematic diagram of the equipment is shown in figure 5.1. The equipment operates in a vacuum chamber maintained at a pressure higher than 3×10^{-4} Pa, with two independent evaporation sources heated by two 7 kW electron guns, using a 10 kV high-voltage supply. The alloy vapour condenses on to a temperature-controlled rotating collector, positioned approximately 150 mm above the evaporation sources and 235 mm in diameter. The temperature-controlled rotating substrate rotates at 20 rpm and is kept at 523 K in order to homogenize the deposit composition. Compositions are controlled by changing Al and Fe evaporation rates by changing the electron-beam power supplied to each source.

Deposit structures are determined by x-ray diffraction (XRD), transmission electron microscopy (TEM) and differential scanning calorimetry (DSC). Microhardness measurements are obtained using an indentor load of 0.245 N. Tensile strengths are measured at room temperature in an Instron testing machine, using bulk deposited material with gauge dimensions of 3 mm in width, 0.15–0.2 mm in thickness and 8 mm in length at an initial strain rate of 6.9×10^{-4} s^{-1}. Fracture morphology is examined by scanning electron microscopy (SEM).

Microstructure

It has been suggested [4] that nonequilibrium phases are formed by rapid quenching. The effect of Fe concentration on phase formation in vapour quenched Al-Fe alloys has been investigated. Figure 5.2 shows x-ray diffraction patterns from $Al_{100-x}Fe_x$ alloy deposits, with $x = 0, 1, 3$ and 5 at%. For the $Al_{95}Fe_5$ alloy, the x-ray diffraction pattern exhibits a broad halo and fcc α-Al peaks, indicating that the deposit consists of a mixed structure of amorphous and α-Al crystalline phases. Diffraction patterns from $Al_{99}Fe_1$

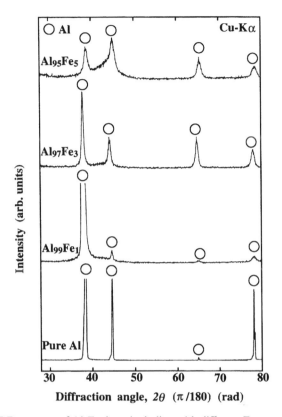

Figure 5.2. XRD patterns of Al-Fe deposited alloy with different Fe contents.

and $Al_{97}Fe_3$ alloys consist of α-Al peaks with no second-phase diffraction peaks, indicating that the Fe additions are in solution in the fcc α-Al. In other words, a supersaturated fcc solid solution phase is achieved in the high Fe content alloys $Al_{99}Fe_1$ and $Al_{97}Fe_3$, with the solubility extended by rapid quenching. The full-width at half maximum (FWHM) of the diffraction peaks increases with increasing Fe content, indicating that grain refinement occurs with increasing addition of Fe. Texture is also affected by the addition of Fe. In pure Al, the grains are oriented randomly, but in $Al_{99}Fe_1$ a strong (111) texture is observed, which diminishes with increasing Fe content.

To clarify the morphology and grain size of the materials, TEM bright-field images and selected area diffraction patterns from the $Al_{100-x}Fe_x$ deposited alloys are shown in figure 5.3. The selected-area diffraction pattern taken from a region with a diameter of 1 μm consists of a diffuse halo from the amorphous phase and diffraction rings from the fcc α-Al phase. The fcc α-Al grain size is approximately 40 nm. figure 5.4 shows

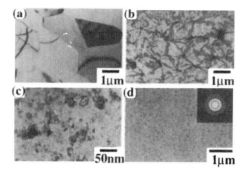

Figure 5.3. TEM bright-field images and diffraction patterns of Al-Fe deposited alloy with different Fe contents.

grain size plotted versus Fe content for the supersaturated solid solution alloys. As shown in figures 5.3 and 5.4, the α-Al grain size decreases significantly from 2000 to 37 nm with increasing Fe content from 1 to 3 at%. This is caused partly by the suppression of grain growth with an ultrahigh cooling rate in electron-beam deposition, and partly by the addition of Fe with a low diffusion coefficient in Al [6]. Overall, addition of Fe is effective at producing grain refinement in electron-beam deposited Al based alloys. Strain contrast caused by the supersaturation of Fe is observed in the interior of the Al-Fe grains, as seen in figures 5.3(b–d), compared with pure Al in figure 5.3(a). It is considered that the strain is introduced by the supersaturation of Fe in α-Al.

Figure 5.5 shows differential scanning calorimeter curves from the $Al_{100-x}Fe_x$ deposited alloys. The $Al_{99}Fe_1$ alloy exhibits a small exothermic peak at about 560 K. X-ray diffraction after heating in the differential scanning calorimeter indicates that this exothermic peak corresponds to precipitation of an Al-Fe intermetallic compound. The $Al_{97.5}Fe2.5$ alloy

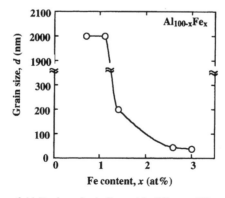

Figure 5.4. Grain size of Al-Fe deposited alloy with different FE contents.

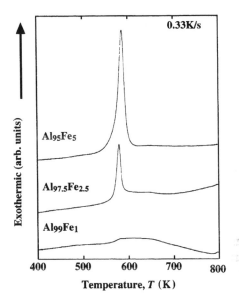

Figure 5.5. DSC curves of Al-Fe deposited alloy with different Fe contents.

also shows an exothermic peak corresponding to precipitation of an Al-Fe intermetallic compound at about 560 K, and the intensity of the precipitation peak increases with increasing Fe content. However, differential scanning calorimeter curves from $Al_{95}Fe_5$ exhibit a sharp intense peak at approximately 570 K. It seems likely that the amorphous phase crystallizes and the supersaturated solid solution decomposes at approximately 570 K.

Mechanical properties

Figure 5.6 shows Vickers hardness of the Al-Fe deposits plotted as a function of Fe content. Two regions are apparent. In the low Fe content region, up to approximately 4 at%, the Vickers hardness increases sharply. The Vickers hardness of the deposits increases to about $200\,H_V$ at an Fe content of 1 at%. Above 4 at%, hardness increases more slowly, and the hardness reaches $600\,H_V$ at an Fe content of 15 at%. The Vickers hardness of a conventional high-strength Al based alloy, Duralumin 7075-T6, is approximately $174\,H_V$. Therefore, very high hardness of is obtained in Al-Fe binary deposits even with dilute Fe additions. The change in the dependence of hardness on composition around 4 at% corresponds to the transition of the microstructure from an fcc α-Al supersaturated solid solution to a two phase mixture of fcc α-Al and an amorphous phase.

The tensile properties of the Al-Fe deposits which consist of supersaturated solid solution are shown in figure 5.7. The tensile strength of the

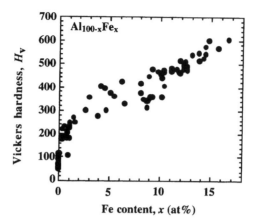

Figure 5.6. Relationship between Fe content and Vickers hardness of Al-Fe deposited alloy.

Al-Fe deposits increases with increasing Fe content and reaches 1000 MPa at 2.5 at% Fe. This trend is similar to that of the composition dependence of hardness. Above 2.5 at%, however, the tensile strength exhibits a low value compared with that at 2.5 at% owing to brittleness and premature fracture during elastic deformation. The tensile properties of the high Fe content alloys (with Fe contents above 4 at% and hardnesses above 400 H_V) were not evaluated due to their brittleness.

Figure 5.8 shows the fracture surfaces of the Al-Fe deposits. Low Fe content deposits consisting of a supersaturated fcc α-Al solid solution show fracture surfaces with well-developed dimple patterns typical of a

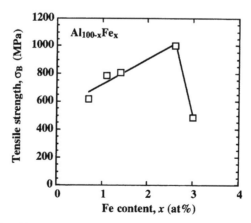

Figure 5.7. Relationship between Fe content and tensile strength of Al-Fe deposited alloy.

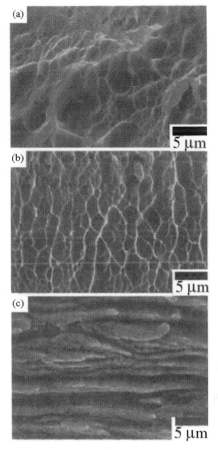

Figure 5.8. Fracture surfaces of Al-Fe deposited alloy with different Fe contents. (a) $Al_{99}Fe_1$, (b) $Al_{97.5}Fe_{2.5}$, (c) $Al_{95}Fe_5$.

ductile material, as shown in figures 5.8(a) and (b). The size of the dimple pattern is smaller at higher Fe contents, indicating that the ductility decreases with increasing Fe content. By contrast, high Fe content deposits consisting of a mixture of amorphous and fcc α-Al phases show typical brittle fracture surfaces, as shown in figure 5.8(c).

Figure 5.9 shows Vickers hardness and tensile strength of the deposits plotted versus the inverse square root of grain size. Straight lines in figure 5.9 indicate that the Al-Fe deposited alloy follows the Hall–Petch relationship with respect to grain size variations between 2000 nm and approximately 40 nm. Therefore, it is concluded that the ultrahigh tensile strength and Vickers hardness of the Al-Fe deposited alloy are achieved by grain refinement which is effective down to approximately 40 nm.

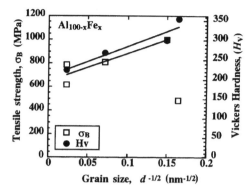

Figure 5.9. Grain size dependence of tensile strength and Vickers hardness of Al-Fe deposited alloy.

Conclusions

Electron-beam evaporated Al-Fe deposits with Fe contents up to 4 at% consist of a supersaturated solid solution of fcc α-Al. At higher Fe contents, the deposits consist of mixture of amorphous and fcc α-Al crystalline phases, with a grain size below 40 nm. In the fcc α-Al supersaturated sold solution range, the deposit grain size decreases drastically from 2000 to 37 nm with increasing Fe content from 1 to 3 at%. Deposit hardness increases with increasing Fe content and reaches values of 200 H_V and 600 H_V for Fe contents of 1 at% and 15 at%, respectively. Deposit tensile strength exhibits a maximum of 1000 MPa at an Fe content of 2.5 at%, where the grain size is 45 nm, with a ductile dimple fracture morphology. Hardness and tensile strength are both proportional to the inverse square root of grain size, indicating that Hall–Petch grain size refinement is the strengthening mechanism, and is effective down to approximately 40 nm.

References

[1] Dodd S B and Morris S 1997 *Society of Vacuum Coaters 40th Annual Technical Conf. Proc.* p 145
[2] Inoue A, Ohtera K and Masumoto T 1988 *Jpn. J. Appl. Phys.* **27** L736
[3] McConnell M C and Partridge P G 1987 *Acta Metall.* **35** 1973
[4] Jones H 1969 *Mater. Sci. Eng.* **5** 1
[5] Willey L A 1967 *Aluminum* **1** 359
[6] Fujikawa S 1996 *J. JLIM* **46** 202

Chapter 6

Electrodeposited nanocrystalline Ni alloys

Tohru Yamasaki

Introduction

The mechanical properties of nanocrystalline materials have been studied extensively over the past two decades because of the expectation that grain size refinement of the materials into the nanometre size range will enhance mechanical properties, such as strength and hardness, possibly without loss of ductility. The possibility of superplastic forming at moderate temperatures and strain rates is also an attractive expectation for such nanocrystalline refinement [1]. However, recent experimental observations have reported that most of the nanocrystalline materials are limited in practical applications because of their brittleness [2]. It has not yet been well understood whether brittle behaviour is an intrinsic feature of nanocrystalline materials, or whether it is caused by processing difficulties such as imperfect consolidation of nanocrystalline powders [3]. The grain size dependence of the strength of nanocrystalline materials in the size range of a few to several tens of nanometres is fundamentally important for understanding the mechanical behaviour of nanocrystalline materials. Experimental results reported so far show considerable scatter. It is, therefore, important to study the mechanical properties using sufficiently dense and structurally well characterized nanocrystalline materials.

Electrodeposition is a good technique for producing materials ranging from amorphous to nanocrystalline with a grain size of about 5–50 nm in bulk form or as a coating, with no post-processing requirements [4]. Electrodeposition can be applied to microfabrication processes such as LIGA (Lithographie, Galvanoformung, Abformung) that requires a hard material. Electrodeposited alloys have high hardness, but are limited in these applications because of their brittleness [5]. For example, amorphous and nanocrystalline Ni-W electrodeposited alloys have excellent properties such as high hardness, high corrosion resistance and high thermal stability. The process for electrodeposition of W-rich Ni-W alloys, however, has not yet been well developed and the electrodeposited alloys are very brittle [6–9]. This chapter describes an aqueous plating bath for Ni-W electrodeposition that yields amorphous and nanocrystalline alloys of fairly high tungsten content, and shows how Ni-W alloys with both high hardness and high ductility can be produced by electrodeposition when the processing conditions are carefully chosen [10, 11].

Chapters 1 and 2 describe the fundamental thermodynamics and structure of nanocrystalline alloys, and chapters 3–5 and 7–9 describe the manufacture and structure of a number of different nanocrystalline metallic materials.

Formation of high-strength Ni-W electrodeposits

Plating bath compositions and conditions are shown in table 6.1. Trisodium citrate and ammonium chloride complexing agents are added to a basic nickel sulphate/sodium tungstate electroplating bath solution to form complexes with nickel and tungsten. The substrate is electropolished copper sheet and the anode is high-purity platinum sheet. The 600 ml plating cell beakers each contain 500 ml of bath solution and are thermostatically controlled to maintain the electroplating temperature. For each deposit, a fresh plating bath is made from analytical reagent grade chemicals and deionized water. The Ni-W alloy deposition rate can be determined by weighing the substrate before and after electrodeposition and calculating the additional mass per square centimetre. Electrodeposited Ni-W films are separated from their substrates by immersion in an aqueous solution containing $250\,g/l$ CrO_3 and $15\,cc/l$ H_2SO_4. The resulting electrodeposited films can then be annealed at various temperatures in vacuum of about $10^{-3}\,Pa$. Ductility is determined by measuring the radius of curvature at which fracture occurs in a simple bending test. The fracture strain on the outer surface of the specimen, ε_f, is estimated from

$$\varepsilon_f = \frac{t}{(2r - t)} \tag{6.1}$$

Table 6.1. Plating bath compositions and conditions for Ni-W electrodeposition.

Nickel sulphate, $NiSO_4 \cdot 6H_2O$	0.06 mol/l
Sodium tungstate, $Na_2WO_4 \cdot 2H_2O$	0.14 mol/l
Trisodium citrate, $Na_3C_6H_5O_7 \cdot 2H_2O$	0.14–0.5 mol/l
Ammonium chloride, NH_4Cl	0.5 mol/l
(or ammonium sulphate, $(NH_4)2SO_4$)	(0.25 mol/l)
Bath temperature	333–363 K
pH	7.5–8.5
Current density	0.05–0.2 A/dm^2

where r is the radius of curvature on the outer surface of the bend sample at fracture and t is the sample thickness. Vickers microhardness is measured on cross sections of the electrodeposited and annealed samples on their copper Cu substrates with a 0.02 kg load and a loading time of 15 s.

Plating bath temperature and applied current density

The tungsten content of the electrodeposits is strongly influenced by the plating bath temperature and the applied current density. Figure 6.1 shows x-ray diffraction patterns from the Ni-W electrodeposits of different tungsten contents for various plating bath temperatures between 333 and 363 K with a trisodium citrate concentration of 0.5 mol/l at an applied current density of $0.2\,A/cm^2$. The tungsten content of the electrodeposits increases with increasing plating bath temperature and an amorphous x-ray pattern appears at tungsten contents of more than about 20 at%.

Figure 6.2 shows differential thermal analysis scans at a heating rate of 0.33 K/s from as-electrodeposited Ni-W alloys for various plating bath temperatures. No distinct crystallization peaks are observed for the Ni-17.8 at% W alloy electrodeposited at a plating bath temperature of 333 K. At a bath temperature of 348 K and above, an amorphous x-ray diffraction pattern appears and crystallization of the amorphous Ni-W alloys takes place in two steps. The first crystallization step at a temperature of about 980 K has been confirmed by x-ray analysis to be due to the formation of an fcc Ni-W solid solution. The second crystallization step takes place at a temperature range between 1100 and 1150 K. X-ray analysis suggests that a Ni_4W intermetallic compound precipitates during this step.

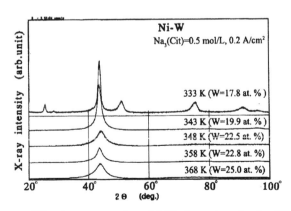

Figure 6.1. X-ray diffraction patterns from Ni-W electrodeposits for different tungsten contents and plating bath temperatures between 333 and 363 K at an applied current density of $0.2\,A/cm^2$ [10].

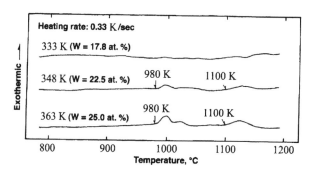

Figure 6.2. DTA measurements at a heating rate of 0.33 K/s for as-electrodeposited Ni-W alloys for various plating bath temperatures [9].

Deposition rate and mechanical properties as a function of plating bath temperature for the Ni-W electrodeposits are shown in table 6.2. The Vickers microhardness of the electrodeposits increases continuously from 602 to 770 with increasing plating bath temperature from 333 to 363 K, while the deposition rate has a maximum value of 68.8 mg/cm^2 h at a plating bath temperature of 348 K and then decreases with increasing plating bath temperature.

Electrodeposited Ni-22.5 at% W at a plating bath temperature of 348 K has a high hardness of $H_V = 685$ and is ductile, i.e. can be bent through an angle of 180° without breaking ($\varepsilon_f = 1.0$). At other plating bath temperatures, the electrodeposits are very brittle. Figure 6.3 shows SEM micrographs of the Ni-22.5 at% W alloy electrodeposited at a plating bath temperature of 348 K after bending through an angle of 180°. The alloy deforms plastically and very inhomogeneously. Shear bands form on the bending edge showing typical features of ductile metallic glasses.

The tungsten content of the Ni-W electrodeposits is also influenced by the applied current density. Figure 6.4 shows x-ray diffraction patterns from

Table 6.2. Deposition rate and mechanical properties versus plating bath temperature for the Ni-W electrodeposits [10].

Bath temperature (K)	W content (at%)	Deposition rate (Mg/cm^2/h)	Vickers hardness (H_V)	Fracture strain (ε_f)	Structure judged by x-ray diffraction
333	17.8	49.2	602	0.0	Nanocrystalline
343	19.9	65.0	650	0.02	Nanocrystalline
348	22.5	68.8	685	1.00 (ductile)	Amorphous
353	22.8	66.7	752	0.416	Amorphous
363	25.0	55.1	770	0.005	Amorphous

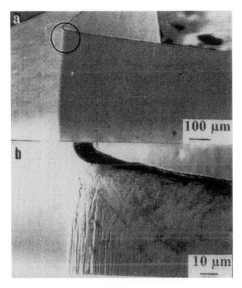

Figure 6.3. SEM micrographs from Ni-22.5at% W electrodeposited at a plating bath temperature of 348 K after bending through an angle of 180° [10].

as-electrodeposited Ni-W alloys for various current densities between 0.05 and 0.2 A/cm² at a plating bath temperature of 348 K. X-ray diffraction peaks from the deposited alloys broaden and the tungsten content increases with increasing applied current density. Under plating conditions with high current density, the Ni-W electrodeposits have amorphous and nanocrystalline structures and are ductile, i.e. they can be bent through an angle of 180° without breaking ($\varepsilon_f = 1.0$).

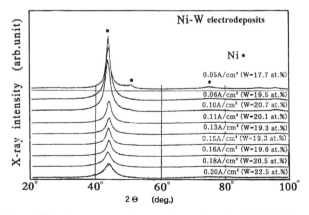

Figure 6.4. X-ray diffraction patterns from as-electrodeposited Ni-W alloys for various current densities between 0.05 and 0.2 A/cm² at a plating bath temperature of 348 K. All the Ni-W alloys are ductile and can be bent through an angle of 180° without breaking [11].

Table 6.3. Average grain sizes and Vickers microhardness of as-deposited Ni-W alloys. Bath temperature: 348 K (723 K for annealing 24 h in vacuum) [10].

Applied current density (A/cm^2)	W content (at%)	Grain size, as-deposited (nm)	Vickers hardness (H_V)	Grain size, 723 K, 24 h (nm)	H_V, 723 K, 24 h
0.05	17.7	6.8 (nanocrystalline)	558 (ductile)	9.5	919 (brittle)
0.10	20.7	4.7 (nanocrystalline)	635 (ductile)	9.0	962 (brittle)
0.15	19.3	4.7 (nanocrystalline)	678 (ductile)	8.9	992 (brittle)
0.20	22.5	<2.5 (amorphous)	685 (ductile)	8.2	997 (brittle)

Average grain sizes and Vickers microhardnesses of as-electrodeposited Ni-W alloys for various current densities between 0.05 and 0.2 A/cm^2 are collected in table 6.3. Average grain sizes in the Ni-W electrodeposits are obtained by applying the Scherrer formula to the diffraction lines of fcc Ni(111) and the broad maximum of the amorphous phase. With increasing applied current density, the average grain size decreases and the Vickers microhardness increases continuously. On annealing these materials at 723 K for 24 h, grain growth occurs, the grain size increases to 8.2–9.5 nm, and the hardness increases significantly to more than $H_V = 900$.

Co-deposited hydrogen and oxygen

As shown in table 6.2, electrodeposited Ni-W alloys at a plating bath temperature of 348 K are ductile. At other plating bath temperatures, the electrodeposits are very brittle. This temperature dependence of the ductility may be due to the effects of inclusion of impurities in the Ni-W electrodeposits such as codeposited hydrogen and oxygen during the deposition process. Figure 6.5 shows the bath temperature dependence of amounts of codeposited hydrogen and oxygen in the Ni-W electrodeposits with various concentrations of trisodium citrate Na$_3$(Cit) in the plating bath solution. For Na$_3$(Cit) = 0.5 mol/l, codeposited hydrogen and oxygen are at their minimum values of 0.0006 and 0.028 at% respectively at a plating bath temperature of 348 K. This minimum temperature decreases with decreasing Na$_3$(Cit) concentration in the plating bath solution. Particularly ductile amorphous and nanocrystalline Ni-W alloys have been obtained at plating bath temperatures between 323 and 348 K. As shown in table 6.4, some of the Ni-W electrodeposits with high hardness can be bent through an angle of 180° without breaking ($\varepsilon_f = 1.0$).

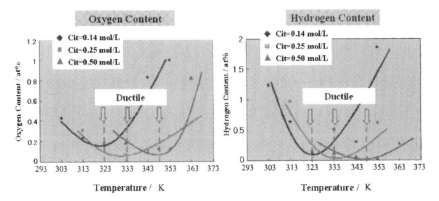

Figure 6.5. Bath temperature dependence of codeposited hydrogen and oxygen in Ni-W electrodeposits with various concentrations of trisodium citrate $Na_3(Cit)$ in the plating bath solution.

TEM and EXAFS structure analysis

Figures 6.6 and 6.7 show low and high magnification transmission electron microscope (TEM) images and corresponding selected area diffraction patterns from as-electrodeposited Ni-12.3 at% W and Ni-20.7 at% W alloys with both high hardness and high ductility. The Ni-12.3 at% W alloy in figure 6.6 has a grain size of about 5 nm and the selected area diffraction patterns consist of fcc Ni Debye rings. The Ni-20.7 at% W alloy in figure 6.7 has a lower grain size of about 3 nm with no clear interface in the 1–2 nm thick intercrystalline grain boundary regions. In the intercrystalline regions, distorted amorphous-like lattice images are observed, and selected area diffraction patterns consist of broad amorphous-like Debye rings.

Nasu and his collaborators [12] have analysed the local structure of the Ni-16.0 at% W and the Ni-20.7 at% W electrodeposited Ni-W alloys having high hardness and high ductility by using extended x-ray analysis of fine spectra (EXAFS) measurements made at beam line 12C of the Photon Factory in KEK, Japan (2.5 GeV storage ring; maximum ring current of 200 mA) at 20 K. Energy scans are made near the Ni-K and W L_{III} edges using a Si(111) monochromator in transmission mode. Figure 6.8 shows the resulting radial distribution functions (RDFs) around the Ni atoms for the electrodeposited Ni-W alloys obtained by Fourier transforming the EXAFS spectra (0.36–1.75 nm^{-1} for the Ni K-edge). The arrows indicate the locations of first, second, third and fourth nearest neighbours around the Ni atoms in crystalline nickel. The radial distribution functions from the electrodeposited Ni-W alloys are very similar to those from crystalline Ni and Ni_4W. The peak positions in the radial distribution functions derived from the measured EXAFS spectra are closer to those from the Ni_4W crystal compound than

Table 6.4. Plating conditions and various properties of the electrodeposited Ni-W alloys.

Composition (at%)	Bath temperature (K)	$Na_3(Cit)$ (mol/l)	Current density (A/cm^2)	Hardness (H_V)	Fracture strain (ε_f)	Structure by x-ray diffraction
Ni-25.0% W	363	0.5	0.2	770	0.0002	Amorphous
Ni-22.5% W	348	0.5	0.2	685	1.0 (ductile)	Amorphous
Ni-20.7% W	348	0.5	0.1	635	1.0 (ductile)	3.0 nm
Ni-17.7% W	348	0.5	0.05	558	1.0 (ductile)	6.8 nm
Ni-12.3% W	323	0.14	0.05	696	1.0 (ductile)	5.2 nm
Ni-10.6% W	313	0.14	0.05	702	0.4	10.6 nm
Ni-9.7% W	303	0.14	0.05	695	0.02	5.4 nm + oxide

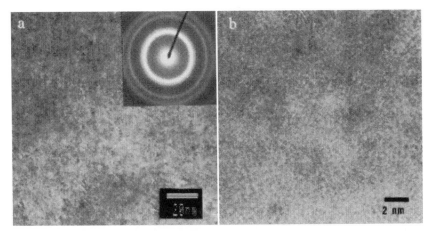

Figure 6.6. Transmission electron micrographs and selected area diffraction pattern from as-electrodeposited Ni-12.3 at% W: (a) low and (b) high magnification.

to those from fcc Ni. The Ni_4W crystal structure is a combination of the structures of body-centred tetrahedral tungsten ($a = 0.573\,nm$, $c = 0.355\,nm$) and face-centred tetrahedral nickel ($a = 0.362\,nm$, $c = 0.355\,nm$) as shown schematically in figure 6.9 [13]. This is why the radial distribution functions around the Ni atoms in the electrodeposited Ni-W alloys are very similar to those from crystalline Ni and Ni_4W. The intensity of all peaks in the radial distribution functions of the Ni-W alloys decreases, and the deviations of the distances of atomic pairs increases with increasing tungsten concentration. However, the alloy still has medium range order and maintains a crystal structure at a tungsten

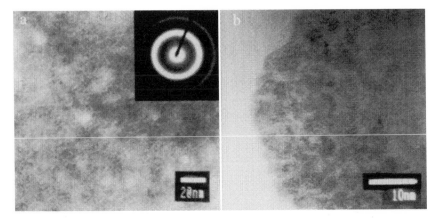

Figure 6.7. Transmission electron micrographs selected area diffraction pattern from as-electrodeposited Ni-20.7 at% W alloy: (a) low and (b) high magnification [11].

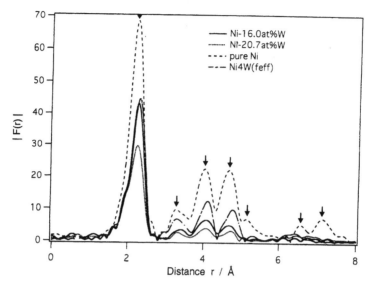

Figure 6.8. RDFs around Ni atoms for electrodeposited Ni-W alloys obtained by Fourier transforming EXAFS spectra (0.36–1.75 nm^{-1} for the Ni K-edge) [12].

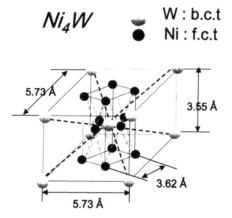

Figure 6.9. Schematic diagram of Ni$_4$W crystal structure, a combination of body centred tetrahedral tungsten ($a = 0.573$ nm, $c = 0.355$ nm) and face-centred tetrahedral nickel ($a = 0.362$ nm, $c = 0.355$ nm).

content of 20.7 at%. Figure 6.10 shows the radial distribution functions around the W atoms for the electrodeposited Ni-W alloys. There is no peak in the radial distribution functions that corresponds to medium range order around the W atoms, indicating considerable deviations in the atomic distances around the W atoms.

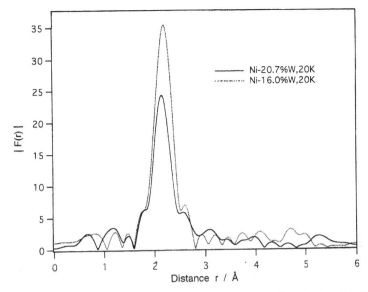

Figure 6.10. RDFs around W atoms for electrodeposited Ni-W alloys obtained by Fourier transforming EXAFS spectra (0.36–1.60 nm^{-1} for the W L_{III}-edge) [12].

Tensile behaviour

Figure 6.11 shows stress–strain curves obtained by tensile testing the electrodeposited Ni-20.7 at% W and Ni-12.3 at% W alloys with grain sizes of about 3–5 nm. In order to clarify the effect of inclusion of codeposited hydrogen during deposition, stress–strain curves for degassed specimens after 24 h at 353 K in vacuum are also shown. The results vary considerably from specimen to specimen, being very sensitive to the residual hydrogen. In the case of the as-electrodeposited Ni-20.7 at% W alloys, the tensile strength and elongation to fracture were 670 MPa and about 0%, respectively. After degassing at 353 K, the tensile strength and elongation to fracture increase significantly, to maximum values of 2333 MPa and about 0.5%, respectively. In the case of the as-deposited Ni-12.3 at% W alloy, the tensile strength and elongation to fracture are 1416 MPa and about 0.2%, respectively. After degassing at 353 K, the tensile strength decreases somewhat to 800–1000 MPa.

Figure 6.12 shows scanning electron micrograph images of fractured surfaces of tensile specimens of the electrodeposited Ni-20.7 at% W and Ni-12.3 at% W alloys after degassing at 353 K for 24 h. Their fracture surfaces are fairly typical of ductile amorphous alloys. Typical river or vein patterns on the fracture surfaces are observed and demonstrate local plastic deformation.

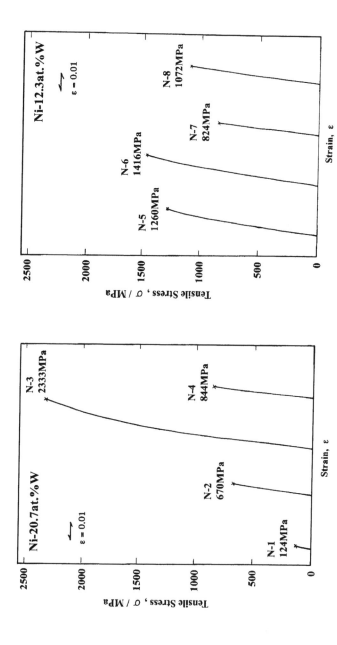

Figure 6.11. Stress–strain curves obtained by tensile testing Ni-20.7 at% W and Ni- 12.3 at% W alloys electrodeposited at plating bath temperatures of 348 K and 323 K respectively. (N-1) and (N-2): as-electrodeposited Ni-20.7 at% W; (N-3) and (N-4): degassed Ni-20.7 at% W at 80 °C for 24 h; (N-5) and (N-6): as-electrodeposited Ni-12.3 at% W; (N-7) and (N-8): degassed Ni-12.3 at% W at 80 °C for 24 h [11].

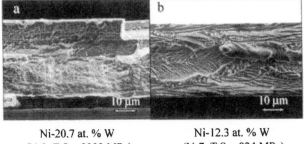

Ni-20.7 at. % W
(N-3, T.S.= 2333 MPa)

Ni-12.3 at. % W
(N-7, T.S.= 824 MPa)

Figure 6.12. SEM images of fractured surfaces of tensile specimens of degassed Ni-W electrodeposited alloys.

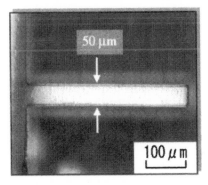

Figure 6.13. Optical micrograph of micro-electroformed high-strength nanocrystalline Ni-20.7 at% W.

Figure 6.13 shows an example of micro-electroforming of the high-strength nanocrystalline Ni-20.7 at% W alloy by using a ultraviolet–lithographic technique. As shown in figure 6.13, high precision forming in the micrometre-size range is possible by Ni-W electrodeposition, indicating that this alloy system may be useful for applications in microfabrication.

Critical grain size for high strength and ductility

As shown in table 6.4, nanocrystalline Ni-W alloys with microstructures ranging from amorphous to nanocrystalline with a grain size of about 7 nm in diameter exhibit high hardness combined with good ductility. Especially, a Ni-20.7 at% W alloy with a grain size of about 3 nm exhibits a high tensile strength of about 2300 MPa, as shown in figure 6.11. However, on annealing these materials, grain growth occurs and they lose their ductility. Figure 6.14 shows transmission electron micrographs and

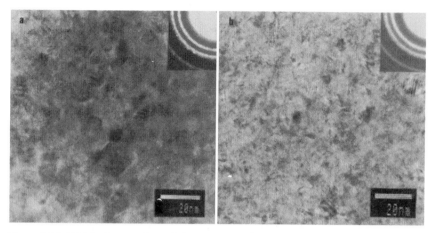

Figure 6.14. TEM micrographs and the corresponding selected area diffraction patterns from Ni-20.7 at% W and Ni-12.3 at% W alloys after annealing at 723 K for 2 h in vacuum.

corresponding selected area diffraction patterns from Ni-20.7 at% W and Ni-12.3 at% W alloys after annealing at 723 K for 2 h in vacuum. Grain growth occurs, grain sizes increase to about 10 nm for both Ni-W alloys, and they lose their ductility.

Figure 6.15 shows fracture strain, ε_f, and Vickers microhardness as a function of grain size for electrodeposited Ni-20.7 at% W and Ni-12.3 at% W alloys. Average grain sizes in the Ni-W electrodeposits are obtained by applying the Scherrer formula to the x-ray diffraction lines of fcc (111) peaks and also by direct observations from transmission electron micrographs. When the grain size increases to about 9 nm, both Ni-W alloys exhibit severe brittleness and the hardness increases to maximum values of about $H_V = 950$ and $H_V = 850$, respectively.

Hall–Petch relationship in nanocrystalline Ni-W alloys

Figure 6.16 shows Vickers microhardness versus annealing temperature for Ni-25.0 at% W alloy after annealing at various temperatures. For comparison, the relationships for Ni-20 at% P [14] and pure nickel [15] electrodeposits are also shown. For this temperature range, the hardness of the Ni-W alloy is high compared with the others. The as-deposited Ni-25.0 at% W alloy with an amorphous structure exhibits a high hardness of about $H_V = 770$. On annealing this material at 673 K for 24 h in air or vacuum, the x-ray diffraction profile becomes gradually sharper and the hardness increases to about $H_V = 1100$. At a higher annealing temperature of 873 K for 24 h in vacuum, a nanocrystalline structure with a grain size

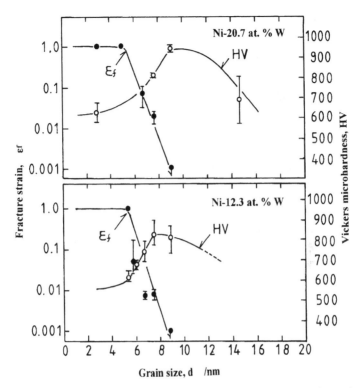

Figure 6.15. Fracture strain, ε_f, obtained by simple bend testing and Vickers microhardness as a function of grain size for Ni-20.7 at% W and Ni-12.3 at% W [11].

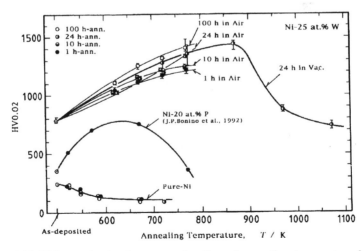

Figure 6.16. Vickers microhardness as a function of annealing temperature for Ni-25.0 at% W, Ni-20 at% P and pure-Ni electrodeposits after various annealing times [9].

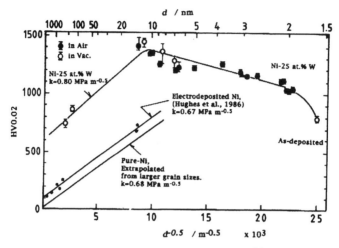

Figure 6.17. Vickers microhardness as a function of grain size$^{-0.5}$ for Ni-25.0 at% W after annealing at various temperatures in air or vacuum [9].

of about 12 nm is observed by x-ray diffraction and the hardness is further increased to a maximum value of $H_V = 1450$. On annealing at 973 K and above, the hardness decreases markedly with increasing grain size.

Figure 6.17 shows Vickers microhardness versus $d^{-0.5}$ for the Ni-W alloy after annealing at various temperatures in air or vacuum. For comparison, similar relationships for electrodeposited Ni [16] and conventional bulk pure Ni extrapolated from results at larger grain sizes [17] are also shown. In the Ni-W alloy, the hardness increases with decreasing grain size to about 10 nm. The Hall–Petch slope of 0.8 MPa m$^{1/2}$ in the Ni-25 at% W alloy is slightly higher, but comparable with those of electrodeposited Ni and conventional bulk Ni. When the grain size is less than about 10 nm, the hardness decreases with decreasing grain size. Recent results have also shown that nanocrystalline pure nickel produced by electrodeposition exhibits Hall–Petch strengthening down to grain sizes near to about 14 nm, but at a finer grain size of 12 nm, the hardness of nanocrystalline Ni decreases in an apparent breakdown of the Hall–Petch relationship [18]. So, such an apparent breakdown of the Hall–Petch relationship may be general, when the grain size is less than about 10 nm.

Figure 6.18 shows transmission electron microscope images and selected area diffraction patterns from the Ni-25.0 at% W alloy annealed at 723 and 873 K for 24 h in vacuum. On annealing at 723 K, as shown in figure 6.18(a), the nanocrystalline structure develops a grain size of between 5 and 8 nm. Noticeably, image contrast from the interface of individual grains is not clearly visible. Selected area diffraction patterns reveal the fcc lattice Debye rings indicating that the ultrafine grains become randomly oriented.

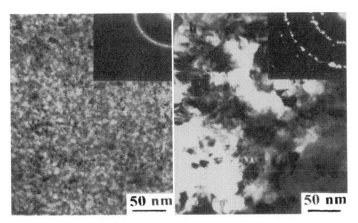

Figure 6.18. TEM images and the selected area diffraction patterns from Ni-25 at% W annealed at (a) 723 K, and (b) 873 K for 24 h in vacuum [9].

On annealing at 873 K, as shown in figure 6.18(b), the grain size increases to about 15 nm and above, and image contrast from the interface of individual grains is clearly visible.

Figure 6.19 shows a high-resolution lattice image of the Ni-W alloy annealed at 723 K for 24 h in vacuum. Individual randomly oriented crystalline grains with grain sizes between 5 and 8 nm are observed. Figure 6.19 corresponds to a defocus value of −48 nm. Distorted lattice images are observed in the 1–2 nm wide intercrystalline regions. As shown in figure 6.20, manually traces of the lattice images in the framed part of figure 6.19 reveal the distorted lattice images in the intercrystalline regions, with

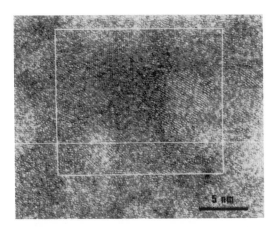

Figure 6.19. High resolution TEM-lattice image of a Ni-W alloy annealed at 723 K for 24 h in vacuum [9].

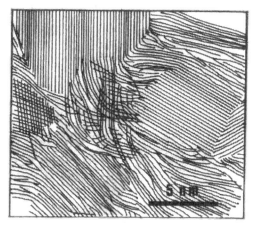

Figure 6.20. Manual tracing of the framed part of the lattice image in figure 6.22 [9].

components of grain boundary and triple junctions (i.e. intersections of two and three adjoining grains). Especially, highly distorted lattice images are observed in the triple junction regions. Straight line lattice images are not observed in the intercrystalline regions when the defocus value is changed from −48 nm to +16 nm. Similar features of high-resolution transmission electron microscope observations have been reported in Ti-Mo alloys prepared by mechanical alloying and Ni_3Al by magnetron sputtering with grain sizes of less than 10 nm [19, 20].

Hardness of nanocrystalline Ni-W alloys

As shown in figure 6.17, Hall–Petch strengthening is observed with the hardness increasing as the grain size decreases to about 10 nm, but the hardness then decreases for grain sizes below about 10 nm. This indicates that a grain size of about 10 nm is the critical one for the mechanical properties of nanocrystalline Ni-W alloys. Figure 6.21 shows the Vickers microhardness versus grain size$^{-.05}$ for a Ni-25.0 at% W alloy.

Fujita has investigated the atomic structure of nanometre-scale crystallites by calculating the volume free energy surface energy of atomic clusters [23]. The interfaces of the nanocrystallites have a non-periodic atomic structure below a critical size. Van Swygenhoven and Caro [24] have also calculated the influence of grain size on the mechanical properties of nanostructured Ni with the grain sizes between 3 and 10 nm by a molecular dynamics computer simulation, and have proposed terms that the grain boundaries give viscoelastic behaviour when the grain sizes fall below 5 nm.

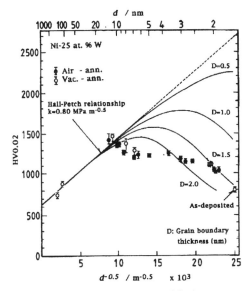

Figure 6.21. Vickers microhardness versus grain size$^{-0.5}$ for Ni- 25.0 at% W [9].

Conclusions

This chapter has discussed the mechanical properties of nanocrystalline alloys in the grain size range up to several tens of nanometres. High-strength nanocrystalline Ni-W alloy with average grain sizes of about 3–5 nm can be manufactured by electrodeposition. The tensile strengths are 1000–2333 MPa combined with good ductility. On annealing at various temperatures, however, grain growth occurs and the materials lose their ductility. Hall–Petch strengthening is observed with the strength and hardness increasing as the grain size decreases to about 10 nm. However, Hall–Petch behaviour is not obeyed and the strength and hardness decrease as the grain size falls below about 10 nm.

References

[1] Karch J, Birringer R and Gleiter H 1987 *Nature* **330** 556
[2] Koch C C, Morris D G, Lu K, Inoue A 1999 *Mater. Res. Soc. Bull.* **24** 54
[3] Morris D G 1998 *Mechanical Properties of Nanostructured Materials, Materials Science Foundations* vol 2, p 61 (Zurich: TransTech Publications)
[4] El-Sherik A M and Erb U 1995 *Plating Surface Finishing* **82** 85
[5] Ehrfeld W, Hessel V, Loewe H, Schulz Ch and Webwer L 1997 *Proc. 2nd Int. Conf. Micro Materials 97, Berlin* p 112
[6] Rauscher G, Rogoll V, Baumgaertner M E and Raub Ch J 1993 *Trans. Inst. Metal Finish.* **71** 95

[7] Domnikov L 1964 *Metal Finishing* **62** 68

[8] Vaaler L E and Holt M L 1946 *J. Electrochemical Soc.* **90** 43

[9] Yamasaki T, Schlossmacher P, Ehrlich K and Ogino Y 1998 *NanoStruct. Mater.* **10** 375

[10] Yamasaki T, Tomohira R, Ogino Y, Schlossmacher P and Ehrlich K 2000 *Plating Surface Finishing* **87** 148

[11] Yamasaki T 2001 *Scripta Mater.* **44** 1497

[12] Nasu T, Sakurai M, Kamiyama T, Usuki T, Uemura O and Yamasaki T 2002 *J. Non-Cryst. Solids* **312–314** 319

[13] Villars P and Calvert L D 1991 *Pearson's Handbook: Crystallographic Data for Inter-metallic Phases (*Materials Park, OH: ASM International)

[14] Bonino J P, Pouderoux P, Rossignol C and Roussert A 1992 *Plating Surface Finishing* **79** 62

[15] Graf P, Schneider W and Zimmermann H 1995 Forchungszentrum, Karlsruhe, private communication

[16] Hughes G D, Smith S D, Pande C S, Johnson H R and Armstrong R W 1986 *Scripta Metall.* **20** 93

[17] Lasalmonie A and Strudel J L 1986 *J. Mater. Sci.* **21** 1837

[18] Schul C A, Nieh T G and Yamasaki T 2002 *Scripta Mater.* **46** 735

[19] Lim W Y, Sukedai E, Hida M and Kaneko K 1992 *Mater. Sci. Forum* **88–90** 105

[20] Van Swygenhoven H, Boni P, Paschoud F, Victoria M and Knauss M 1995 *Nano-Struct. Mater.* **6** 739

[21] Palumbo G, Thorpe S J and Aust K T 1990 *Scripta Metall.* **24** 1347

[22] Palumbo G, Erb U and Aust K T 1990 *Scripta Metall.* **24** 2347

[23] Fujita H 1991 *Ultramicroscopy* **39** 369

[24] Van Swygenhoven H and Caro A 1997 *Abstract of Int. Symp. on Metastable, Mechanically Alloyed and Nanocrystalline Materials, ISMANAM–97, Barcelona*

Chapter 7

Ni, Cu and Ti amorphous alloys

Do Hyang Kim

Introduction

Since Duwez and co-workers [1] reported, in the early 1960s, the formation of a metallic glass in the Au-Si system by rapid solidification, there has been considerable progress in understanding rapid quenching technology and alloy development to manufacture metallic glasses. Metallic glass alloys show characteristic physical properties such as high strength, corrosion resistance and electromagnetic properties, which are significantly different from the corresponding crystalline alloys due to the different atomic configuration [2]. Due to the requirement of rapid solidification, however, metallic glasses can only be produced in the form of either thin ribbon or fine powders. This limits the wide application of metallic glasses except for special purposes such as magnetic materials and brazing.

Recently, several multi-component alloys capable of solidifying into a metallic glass at a relatively low cooling rate have been developed. The development of multicomponent alloys which require only very low cooling rates of about 1–10 K/s to vitrify without crystallization permits the production of large-scale bulk metallic glass (BMG) samples with diameters as large as 30 mm [3]. Several bulk amorphous alloys have been developed in Zr [4], Pd [5], Fe [6], Cu [7], Ti [8] and Mg [9] alloy systems.

Bulk metallic glasses are known to have unique mechanical properties, including high strength, relatively low Young's modulus and perfect elastic behaviour [10]. However, they show little overall room temperature plasticity, but rather deform by highly localized shear banding, resulting in catastrophic failure [11]. Recent attempts to improve the ductility of bulk metallic glasses have focused on the preparation of composite materials exhibiting a microstructure consisting of crystalline phases such as particles [12], fibres [13] or *in-situ* formed precipitates [14] dispersed in the metallic glass matrix. These metallic glass matrix composites have been found to exhibit enhanced plasticity, not generally observed in monolithic bulk metallic glasses [15]. The

121

modified deformation behaviour stems possibly from the formation of multiple shear bands initiated at the interface between the reinforcing agent and the metallic glass matrix, and the confinement of the shear bands in the metallic glass matrix region [16]. Improved ductility may open the possibility of overcoming the limited applications of monolithic bulk metallic glasses that fail catastrophically with little plastic elongation in an apparently brittle manner. However, only a few alloy systems with high glass forming ability, for example, Zr-based [17] and Ti-based [18] metallic glasses, have been obtained as bulk metallic glass matrix composites.

Recently, it has been shown that the production of bulk metallic glasses by consolidation of amorphous powders has promising and important practical applications in the near-net-shape fabrication of components with novel properties [19]. Successful consolidation of amorphous powders has been reported in several alloy systems such as Zr-, Cu- and Ni-based systems [19, 22, 23]. Alternatively several alloy systems based on Ti [8, 20], Cu [7] and Ni [4, 21] have a sufficiently high glass forming ability to enable the fabrication of ingots based on a copper mould casting method.

The main aim of this chapter is to review recent progress in alloy design for enhancing glass forming ability in Ni-, Cu- and Ti-based bulk metallic glasses, and in the synthesis of bulk metallic glass matrix composites. Chapters 1 and 2 describe the fundamental thermodynamics and structure of nano-crystalline alloys, and chapters 3–6, 8 and 9 describe the manufacture and structure of a number of different nanocrystalline metallic materials

Ni-based bulk metallic glasses

Even though many Ni-based amorphous alloys have been produced by rapid quenching techniques [24], Ni-based bulk amorphous alloys have been reported only recently by Inoue *et al* [4]. Fully amorphous rods with a maximum diameter of 1 mm have been prepared in the Ni-Nb-Cr-Mo-P-B system [4] containing large additions of P and B ($\sim$20 at% in total). However, the addition of a large amount of P and B brings about economical and technological demerits. For example, because of the high vapour pressure of P, specific facilities are required during the P addition into the alloy melt.

Based on the three empirical rules for having large glass forming ability, the ternary Ni-Ti-Zr alloy system was selected as a candidate non-P and B alloy with high glass forming ability, with a large atomic size difference and strong negative interaction between the constituent elements. The ternary amorphous alloy shows a clear glass transition behaviour during continuous heating, indicating higher glass forming ability than the constituent binary alloys. Further glass forming ability is achieved by addition of Si and Sn. The temperature range of the supercooled liquid region,

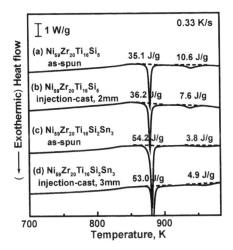

Figure 7.1. Typical DSC spectra obtained during continuous heating at a rate of 0.33 K/s: (a) as-spun $Ni_{59}Zr_{20}Ti_{16}Si_5$; (b) injection-cast $Ni_{59}Zr_{20}Ti_{16}Si_5$; (c) as-spun $Ni_{59}Zr_{20}Ti_{16}Si_5$; and (d) injection-cast $Ni_{59}Zr_{20}Ti_{16}Si_2Sn_3$.

which is often quoted as a parameter representing glass forming ability, increases from 14 K for $Ni_{59}Zr_{20}Ti_{21}$ to 46 K for $Ni_{59}Zr_{20}Ti_{16}Si_5$ and 58 K for $Ni_{59}Zr_{20}Ti_{16}Si_2Sn_3$. Fully amorphous rods with diameters of 2 mm and 3 mm respectively are obtained by injection casting $Ni_{59}Zr_{20}Ti_{16}Si_5$ and $Ni_{59}Zr_{20}Ti_{16}Si_2Sn_3$ alloys [25].

Figure 7.1 shows typical differential scanning calorimetry spectra obtained during continuous heating at a heating rate of 0.33 K/s for melt-spun and injection-cast $Ni_{59}Zr_{20}Ti_{16}Si_5$ and $Ni_{59}Zr_{20}Ti_{16}Si_2Sn_3$ alloys. The injection-cast $Ni_{59}Zr_{20}Ti_{16}Si_5$ and $Ni_{59}Zr_{20}Ti_{16}Si_2Sn_3$ alloys show almost the same crystallization behaviour as the melt spun ribbons. The integrated heats of the exothermic reactions for the bulk specimens are also almost the same as those in the melt-spun specimens.

In order to increase the section size of the bulk amorphous alloys, warm consolidation has been performed for gas atomized amorphous $Ni_{59}Zr_{20}Ti_{16}Si_2Sn_3$ powder [23]. Amorphous 10 mm diameter rods have been successfully fabricated by warm extrusion with an extrusion ratio of 5. The extruded amorphous alloy has a compressive fracture strength of about 2 GPa, which is slightly smaller than 2.2 GPa for an injection-cast rod. No fracture occurs along the prior particle boundaries. This is attributed to good bonding between the particles. The high strength of the material appears to originate from this good particle bonding. Vein patterns, a typical fracture mode characteristic of the amorphous alloys, are clearly observed on the fracture surfaces of the prior powder particles, as can be seen in figure 7.2. These bulk amorphous alloys show excellent corrosion resistance in both acid and sea water [26].

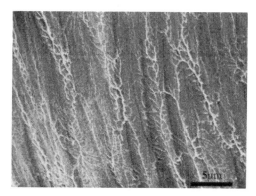

Figure 7.2. Fracture surface of $Ni_{59}Zr_{20}Ti_{16}Si_2Sn_3$ amorphous alloy fabricated by warm pressing.

Cu-based bulk metallic glasses

Cu-Ti-Zr-Ni alloys have been reported to show a large undercooled liquid region before crystallization and a high glass-forming ability [7]. Bulk amorphous specimens with a thickness of 4 mm have been successfully produced by injection casting into a copper mould. The critical cooling rate for amorphous formation is about 250 K/s for this alloy system. A small Si addition to Cu-Ti-Zr-Ni alloys stabilises the undercooled liquid even more against crystallization, and further enhancement of glass forming ability has been obtained by addition of Sn [27]. Figure 7.3 shows a map of the phases for

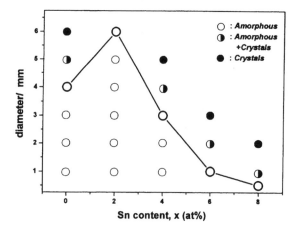

Figure 7.3. Phase map describing the structure of bulk material with Sn content x in $Cu_{47}Ti_{33}Zr_{11}Ni_{8-x}Sn_xSi_1$ ($x = 0, 2, 4, 6, 8$) alloys: ○ amorphous phase; ◑ amorphous and crystalline phase mixture; ● crystalline phase.

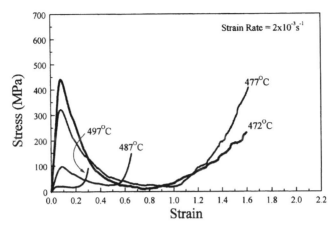

Figure 7.4. Stress–stain curves obtained at several different temperatures between 447–497 °C at a strain rate of 2×10^{-3} s^{-1} for Cu$_{47}$Ti$_{33}$Zr$_{11}$Ni$_6$Sn$_2$Si$_1$ glassy sheet.

a bulk material as a function of rod diameter and alloy composition, determined by x-ray diffraction experiments, with ○, ◑ and ● symbols representing amorphous, mixed amorphous and crystalline, and crystalline phases respectively [28]. Cu$_{47}$Ti$_{33}$Zr$_{11}$Ni$_8$Si$_1$ alloys have a maximum diameter of 4 mm for amorphous formation, smaller than seen previously [27], probably because of a higher oxygen content of about 3000 ppm in wt%. With increasing Sn content x, the maximum diameter increases up to 6 mm at $x = 2$, followed by a decrease. The results clearly show that glass formation can be improved by partial substitution of Ni by Sn. The measured maximum diameter of the amorphous rod is in good correlation with the reduced glass transition temperature T_g/T_m rather than the supercooled liquid region ΔT_x at least in this alloy system.

Figure 7.4 shows flow stress–strain curves obtained at several different temperatures in the supercooled liquid region [29]. With increasing testing temperature, the flow stress decreases as expected. The flow stress reaches a peak just after yielding and then decreases significantly with increasing strain. After the flow stress reaches the minimum plateau level, it rises again before failure. It has been confirmed that deformation in the supercooled liquid state leads to an acceleration of the crystallization kinetics.

Ti-based bulk metallic glasses

Large values of supercooled liquid region ΔT_x, >50 K, have been reported in Ti-Cu-Ni-Sn amorphous alloys. Zhang and Inoue [30] reported a large value of $\Delta T_x = 60$ K in the amorphous Ti$_{50}$Cu$_{25}$Ni$_{20}$Sn$_5$ alloy. Kim *et al* [8] further extended the supercooled liquid region to $\Delta T_x = 73$ K in the amorphous

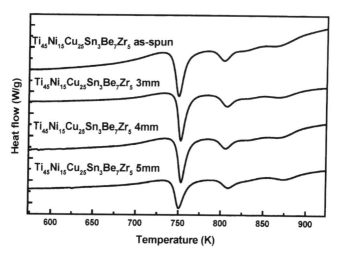

Figure 7.5. DSC curves from melt-spun and copper mould cast $Ti_{45}Ni_{15}Cu_{25}Sn_3Be_7Zr_5$ amorphous alloys.

$Ti_{50}Cu_{32}Ni_{15}Sn_3$ alloy. Further extension of glass forming ability has been obtained by addition of Be [31]. Be can improve the glass forming ability because of its small atomic size and because of the strong interaction between Ti and Be.

A cast 2 mm diameter rod of $Ti_{50}Cu_{32}Ni_{15}Sn_3$ consists of an amorphous phase containing a small amount of crystalline phase. For $Ti_{50}Cu_{25}Ni_{15}Sn_3Be_7$ and $Ti_{45}Cu_{25}Ni_{15}Sn_3Be_7Zr_5$ alloys, however, fully amorphous rods with diameters of 2 and 4 mm, respectively, have been successfully obtained. Figure 7.5 shows differential scanning calorimetry traces obtained from injection cast bulk specimens of $Ti_{45}Cu_{25}Ni_{15}Sn_3Be_7Zr_5$ alloy bars with different diameters, together with equivalent results from melt-spun ribbons. The alloy bars with diameters of 3 and 4 mm show almost the same crystallization behaviour and the same exothermic heat of crystallization as those of the melt-spun ribbon, indicating the formation of fully amorphous structure in the rod specimens. The compressive fracture strength, Young's modulus and total elongation are 2231 MPa, 53.1 GPa and 4.2%, respectively, for $Ti_{50}Cu_{25}Ni_{15}Sn_3Be_7$, and 2328 MPa, 68.4 GPa and 3.4%, respectively, for $Ti_{45}Cu_{25}Ni_{15}Sn_3Be_7Zr_5$. The Vickers hardness is 670 Hv for $Ti_{50}Cu_{25}Ni_{15}Sn_3Be_7$ and 715 Hv for $Ti_{45}Cu_{25}Ni_{15}Sn_3Be_7Zr_5$. The fracture surface shows a well-developed vein pattern.

Other alloy systems

Besides the multicomponent alloys described above, ternary alloys such as Cu-(Zr,Hf)-Ti [32], Cu-Zr-Al [33], and Ni-Nb-(Sn,Ti,Ta) [34–36] have

Table 7.1. Glass transition temperature T_g, crystallization temperature T_x and critical diameter d for bulk glass formation in Ni-Nb based amorphous alloys.

Alloy system	T_g (K)	T_x (K)	Heating rate (K/s)	Critical diameter d (mm)	Reference
$Ni_{65}Nb_5Cr_5Mo_5P_{14}B_6$	703	753	0.67	1	[4]
$Ni_{60}Nb_{20}Ti_{15}Zr_5$	841	898	0.67	2	[56]
$Ni_{60}Nb_{20}Ti_{12.5}Hf_{7.5}$	848	908	0.67	1.5	[57]
$Ni_{53}Nb_{20}Ti_{10}Zr_8Co_6Cu_3$	846	897	0.67	3	[58]
$Ni_{59.35}Nb_{34.45}Sn_{6.2}$	882	930	0.33	3	[34]
$Ni_{60}Nb_{36}Sn_3B_1$	882	940	0.33	3	[34]
$Ni_{60}Nb_{25}Ti_{15}$	857	904	0.67	1.5	[35]
$Ni_{60}Nb_{30}Ta_{10}$	934	961	0.67	2	Present study
$Ni_{60}Nb_{20}Ta_{20}$	*	994	0.67	—	Present study

*Cannot be measured due to the temperature limit in the DSC equipment.

been reported to have high enough glass forming ability to form bulk metallic glasses. These simple ternary alloys can be more suitable for commercial use, and provide a good opportunity for the study of glass-forming mechanisms. Moreover, rapid and simple production techniques are preferable for developing widespread use of bulk metallic glasses in industrial products.

In the case of Ni-Nb-Ta alloys a fully amorphous rod with a diameter of 2 mm can be fabricated by injection casting a $Ni_{60}Nb_{30}Ta_{10}$ alloy. The compressive failure strength of the $Ni_{60}Nb_{30}Ta_{10}$ bulk amorphous alloy is 3346 MPa. As shown in table 7.1, Ni-Nb-Ta metallic glasses exhibit the highest level of glass transition and crystallization temperatures among the Ni-based metallic glasses reported so far. Since the melting range shifts to a higher temperature range with increasing Ta content, the improvement of glass forming ability can be explained by the significantly improved thermal stability of the amorphous phase against crystallization when Nb is partly substituted by Ta.

Recently, new Ca-Mg-Zn and Mg-Cu-Gd alloys having significantly improved glass forming ability have been developed [37, 38]. Ternary $Ca_{65}Mg_{15}Zn_{20}$ bulk metallic glass with a diameter of at least 15 mm can be successfully fabricated by conventional copper mould casting in air. The critical cooling rate for glass formation in the cone-shaped copper mould is less than 20 K/s.

Bulk metallic glass matrix composites

One way to provide substantial ductility in a bulk metallic glass is to produce a composite microstructure, consisting of crystalline and amorphous phases.

The crystalline phase distributed in the amorphous matrix can act as an initiation site for shear bands, resulting in multiple shear band formation throughout the material [39–42]. One effective way to obtain a composite microstructure is to precipitate a nanoscale crystalline phase through partial crystallization of the bulk metallic glass. Unfortunately structural relaxation or extensive precipitation of a crystalline phase from the amorphous matrix during annealing can often make the material more brittle than the as-cast amorphous alloy.

Precipitation of a quasicrystalline phase from the amorphous matrix during annealing has been reported in many alloy systems, such as Pd-U-Si [43, 44], Al-based [45, 46], Ti-based [47] and Zr-based [48, 49] alloys. In particular, formation of quasicrystalline phases in bulk metallic glass forming alloys has recently attracted significant interest, because the strength and ductility of the bulk metallic glass can be enhanced by precipitation of a nanoscale quasicrystalline phase in the amorphous matrix. An improvement in the mechanical properties by precipitation of a quasicrystalline phase has been reported in Zr-based bulk metallic glass forming alloy systems such as Zr-Al-Ni-Cu-Pd [50, 51], Zr-Al-Ni-Cu-Ag [50] and Zr-Ti-Cu-Ni-Al [52]. Recently, a stable icosahedral phase has been shown to precipitate on annealing Ti-rich Ti-Zr-Be-Cu-Ni bulk metallic glass [53]. Table 7.2 shows the effect of quasicrystalline phase precipitation during annealing of the amorphous phase on the mechanical behaviour of bulk glass forming Ti-Zr-Be-Cu-Ni alloys, indicating clearly that *in-situ* composites in which nanosized quasicrystals are isolated and homogeneously distributed in an amorphous matrix exhibit simultaneous improvements in strength and ductility [54].

The simultaneous increase of the strength and compressive strain by partial crystallization of an amorphous phase indicates that there is an increase in the number of shear bands formed during deformation. The nanosized quasicrystalline particles can act as nucleation sites for shear band formation and barriers to shear band propagation, leading to increases

Table 7.2. Mechanical properties of as-cast amorphous and partially crystallized $Ti_{40}Zr_{29}Cu_8Ni_7Be_{16}$ alloy rods: Young's modulus E, yield stress σ_y, ultimate compression stress σ_{max}, yield strain ε_y and fracture strain ε_f.

Fraction of quasicrystal (%)	E (GPa)	σ_y (MPa)	σ_{max} (MPa)	ε_y (%)	ε_f (%)
0	96	1893	1921	2.0	5.1
7	106	2038	2084	1.9	6.2
21	101	1911	2005	1.9	3.4
35	131	1978	2026	1.5	1.7
70	139	1876	1876	1.4	1.4

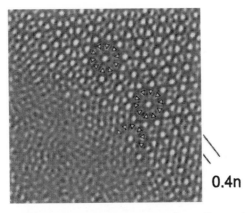

Figure 7.6. High resolution TEM image of the quasicrystal/amorphous interface in a partially quasicrystallized $Ti_{40}Zr_{29}Cu_8Ni_{17}Be_{16}$ alloy.

in yield and fracture strengths. The nature of the interface is an important factor in determining the mechanical properties of the composite alloys. The high interfacial strength between the crystalline phase and amorphous matrix can be effective in preventing shear band propagation without crack initiation, and allows the interfaces to act as sources of multiple shear bands. A Fourier filtered high-res olution image showing the atomic structure of the quasicrystal/amorphous interface in partially crystallized Ti-Zr-Be-Cu-Ni bulk metallic glass (figure 7.6) clearly indicates a gradual structural change from the quasicrystalline to amorphous structure at the interface. With considerable structural similarity between the quasicrystalline and amorphous phases, the interfacial energy between is expected to be lower than that between the crystalline and amorphous phases, leading to a simultaneous increase in strength and ductility.

The ductility of the bulk amorphous alloys can be improved by incorporating a ductile phase in the amorphous matrix. Figure 7.7 shows the microstructure of a ductile brass phase dispersed in a $Ni_{59}Zr_{20}Ti_{16}Si_2Sn_3$ amorphous matrix in a composite fabricated by warm extrusion [55]. Figure 7.8 shows stress–strain curves obtained from the ductile brass phase dispersed bulk metallic glass matrix composites, indicating that catastrophic failure can be avoided by introducing a ductile second phase in the metallic glass matrix. As shown in figure 7.9, many shear bands initiate from the elongated brass powder particles during compressive loading. There is a clear tendency for the shear bands to remain confined between adjacent brass particles, rather than to pass through them. Upon yielding of the bulk metallic glass matrix composites, the soft brass phase deforms plastically at first, and then load is transferred to the surrounding glass matrix, causing the initiation of shear bands at several places of the brass/matrix

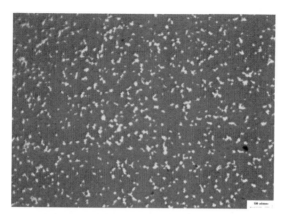

Figure 7.7. Optical micrograph of a bulk metallic glass composite containing 20 vol% brass. The brass powder is well distributed in the metallic glass matrix.

interface. The initiated shear bands propagate and continue to interact with neighbouring brass powder particles. The propagation of single shear bands is found to be restricted to the volume between the brass powder particles, and further propagation through the brass powder is prevented. This behavior is quite similar to that of finely dispersed *in-situ* ductile phase reinforced bulk metallic glass composites. Because of this mechanism of initiating and confining shear bands, bulk metallic glass matrix composites do not fail catastrophically by a single shear band propagating through the whole monolithic sample. Instead, plasticity is distributed more homogeneously in the form of shear band patterns, and this results in higher strains to failure.

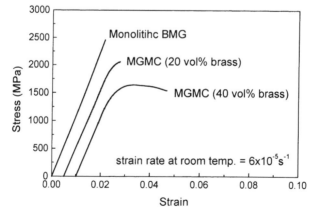

Figure 7.8. Stress–strain curves from a bulk metallic glass and corresponding composites containing 20 and 40 vol% brass under uniaxial compressive testing conditions at room temperature. The load is applied along the extrusion direction.

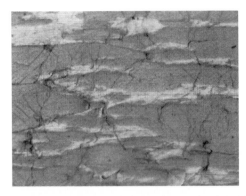

Figure 7.9. Optical images taken from the outer surface of a failed bulk metallic glass composite containing 40 vol% brass. Many shear bands are initiated from the interface between the brass particles and the matrix, but they do not propagate through neighbouring brass particles.

Summary

Significant progress has been made in enhancing the glass forming ability of Ni-, Cu- and Ti-based alloys. However, the maximum size of the amorphous alloy is still relatively small for engineering applications. Further increases in the material dimensions are possible through warm processing of amorphous alloy powders in the wide supercooled liquid region. Improvement of ductility by successful synthesis of *in-situ* and *ex-situ* bulk metallic glass matrix composites suggests that high strength, corrosion resistant bulk amorphous materials can be used in the near future.

References

[1] Klement W, Willens R H and Duwez P 1960 *Nature* **187** 869

[2] Inoue A 2000 *Acta Mater.* **48** 279

[3] Inoue A and Zhang T 1996 *Mater. Trans. JIM* **37** 185

[4] Wang X, Yoshii I, Inoue A, Kim Y H and Kim I B 1999 *Mater. Trans. JIM* **40** 1130

[5] Inoue A, Nishiyama N and Matsuda T 1996 *Mater. Trans. JIM* **37** 181

[6] Lim H K, Yi S, Kim W T, Kim D H, Kim S H and Kim N J 2001 *Scripta Mater.* **44** 1635

[7] Lin X H and Johnson W L 1995 *J. Appl. Phys.* **78** 6514

[8] Kim Y C, Yi S, Kim W T and Kim D H 2001 *Mater. Sci. Forum* **360–362** 67

[9] Kang H G, Park E S, Kim W T, Kim D H and Cho H K 2000 *Mater. Trans. JIM* **41** 846

[10] He G, Eckert J and Löser W 2003 *Acta Mater.* **51** 1630

[11] He G, Eckert J, Löser W and Schultz L 2003 *Nature Mater.* **2** 33

[12] Choi-Yim H, Busch R, Köster U and Johnson W L 1999 *Acta Mater.* **47** 2455
[13] Kim C P, Bush R, Masuhr A, Choi-Yim H and Johnson W L 2001 *Appl. Phys. Lett.* **79** 1456
[14] Fan C, Ott R T and Hufnagel T C 2002 *Appl. Phys. Lett.* **81** 1020
[15] He G, Löser W and Eckert J 2003 *Scripta Mater.* **48** 1531
[16] Hays C C, Kim C P and Johnson W L 2000 *Phys. Rev. Lett.* **84** 2901
[17] Szuecs F, Kim C P and Johnson W L 2001 *Acta Mater.* **49** 1507
[18] He G, Löser W, Eckert J and Schultz L 2002 *J. Mater. Res.* **17** 3015
[19] Kawamura Y, Kato H, Inoue A and Masumoto T 1997 *Int. J. Powder Met.* **30** 50
[20] Zhang T and Inoue A 1999 *Mater. Trans. JIM* **40** 301
[21] Yi S, Park T G and Kim D H 2000 *J. Mater. Res.* **15** 2429
[22] Sordelet D J, Rozhkova E, Huang P, Wheelock P B, Besser M F, Kramer M J, Calvo-Dahlborg M and Dalhborg U 2002 *J. Mater. Res.* **17** 186
[23] Lee M H, Bae D H, Kim W T, Kim D H and Sordelet D J 2003 *J. Non-Cryst. Solids* **315** 89
[24] Park T G, Yi S and Kim D H 2000 *Scripta Mater.* **43** 109
[25] Lee J K, Yi S, Kim W T and Kim D H 2002 *J. Non-Cryst. Solids* submitted for publication
[26] Pang S J, Zhang T, Asami K and Inoue A 2002 *Acta Mater.* **50** 489
[27] Choi-Yim H, Bush R and Johnson W L 1998 *J. Appl. Phys.* **83** 7993
[28] Park E S, Lim H K, Kim W T and Kim D H 2002 *J. Non-Cryst. Solids* **298** 15
[29] Bae D H, Lim H K, Kim S H, Kim D H and Kim W T 2002 *Acta Mater.* **50** 1749
[30] Zhang T and Inoue A 1998 *Mater. Trans. JIM* **39** 1001
[31] Kim Y C, Kim W T and Kim D H 2002 *Mater. Tran. JIM* **43** 1243
[32] Inoue A, Zhang W, Zhang T and Kurosaka K 2001 *Acta Mater.* **49** 2645
[33] Inoue A and Zhang W 2002 *Mater. Trans. JIM* **43** 2921
[34] Choi-Yim H, Xu D H and Johnson W L 2003 *Appl. Phys. Lett.* **82** 1030
[35] Zhang W and Inoue A 2002 *Mater. Trans. JIM* **43** 2342
[36] Lee M H, Bae D H, Kim W T and Kim D H 2003 *Mater. Trans. JIM* **44**
[37] Park E S and Kim D H 2003 *J. Mater. Res.* submitted for publication
[38] Men H and Kim D H 2003 *J. Mater. Res.* **18** 1502
[39] Eckert J, Reger-Leonhard A, Weiß B and Heilmaier M 2001 *Mater. Sci. Eng. A* **301** 1
[40] Inoue A, Zhang T, Chen M W and Sakurai T 2000 *J. Mater. Res.* **15** 2195
[41] Conner R D, Dandliker R B and Johnson W L 1998 *Acta Mater.* **46** 6189
[42] Choi-Yim H and Johnson W L 1997 *Appl. Phys. Lett.* **71** 3808
[43] Poon S J, Drehman A J and Lawless K R 1985 *Phys. Rev. Lett.* **55** 2324
[44] Chen Y, Poon S J and Shiflet G J 1986 *Phys. Rev. B* **34** 3516
[45] Tsai A P, Inoue A, Bizen Y and Masumoto T 1989 *Acta Metall.* **37** 1443
[46] Holzer J C and Kelton K F 1991 *Acta Metall.* **39** 1833
[47] Molokanov V V and Chebotnikov V N 1990 *J. Non-Cryst. Solids* **117/118** 789
[48] Köster U, Meinhardt J, Roos S and Liebertz H 1996 *Appl. Phys. Lett.* **69** 179
[49] Lee J K, Choi G, Kim W T and Kim D H 2000 *Appl. Phys. Lett.* **77** 978
[50] Inoue A, Zhang T, Saida J and Matsushita M 2000 *Mater. Trans. JIM* **41** 1511
[51] Inoue A, Fan C, Saida J and Zhang T 2000 *Sci. Tech. Adv. Mater.* **1** 73
[52] Xing L Q, Eckert J and Schultz L 1999 *NanoStruct. Mater.* **12** 503
[53] Kim Y C, Park J M, Lee J K, Kim W T and Kim D H 2003 *Mater. Trans. JIM* **44**
[54] Kim Y C, Na J H, Park J M, Kim D H and Kim W T 2003 *Appl. Phys. Lett.*

[55] Bae D H, Lee M H, Kim D H and Sordelet D J 2003 *Appl. Phys. Lett.* **83** 2312
[56] Inoue A, Zhang W and Zhang T 2002 *Mater. Trans. JIM* **43** 1952
[57] Zhang W and Inoue A 2003 *Scripta Mater.* **48** 641
[58] Zhang T and Inoue A 2002 *Mater. Trans. JIM* **43** 708

Chapter 8

Nanoquasicrystallization in Zr alloys

Eiichiro Matsubara and Takahiro Nakamura

Introduction

Inoue *et al* [1] have found bulk metallic glasses in Zr-Al-TM (TM = Mn, Fe, Co, Ni and Cu) systems that include neither noble metals nor lanthanide metals. Recently, in some multi-component Zr-based alloys, formation of an icosahedral quasicrystalline phase has been reported [2, 3], which is sensitively affected by cooling rate and oxygen content [4, 5]. Chen *et al* [6] found that addition of Ag improves the formation of the quasicrystalline phase in amorphous Zr-based alloys. Reproducible formation of the quasicrystalline phase has been confirmed in amorphous Zr-Al-Ni-M (M = Ag, Pd, Au and Pt) [7, 8]. An icosahedral quasicrystalline phase was confirmed as the primary precipitation phase in melt-spun $Zr_{70}Ni_{10}M_{20}$ (M = Pd, Au and Pt) ternary metallic glasses with a two-stage crystallization process [9]. Saida *et al* also found a nanometre-sized icosahedral quasicrystalline phase even in a rapidly solidified $Zr_{80}Pd_{20}$ [10] and $Zr_{70}Pd_{30}$ [11] binary alloys. The possible existence of icosahedral short range order in the amorphous state due to large negative heats of mixing for Zr-M pairs has been suggested as important for the formation of nanometre-sized quasicrystals. Anomalous x-ray scattering (AXS) measurements at Zr and Pt absorption edges in an amorphous $Zr_{70}A_{16}Ni_{10}Pt_{14}$ alloy indicates the presence of short range ordered clusters around the Pt atoms [12]. Takagi *et al* [13] also suggested that there are icosahedral-like clusters in amorphous $Zr_{70}Pd_{30}$ alloy based on computer simulation of pair-distribution functions determined by transmission electron microscopy. Saida *et al* [14] observed images of icosahedral clusters in amorphous $Zr_{70}Pd_{30}$ alloy by high-resolution transmission electron microscopy.

This chapter summarizes a series of structural studies in amorphous and quasicrystalline $Zr_{70}Pd_{30}$, $Zr_{80}Pt_{20}$, $Zr_{70}Ni_{10}Pt_{20}$ and $Zr_{70}Ni_{10}Au_{20}$ alloys by anomalous x-ray scattering. Structural characteristics found in the amorphous and quasicrystalline alloys give us clues to understand the mechanism

Table 8.1. Alloy densities.

Samples	g/cm^3
$Zr_{70}Pd_{30}$ amorphous	7.71
$Zr_{70}Pd_{30}$ quasicrystal	7.78
$Zr_{80}Pt_{20}$ quasicrystal	8.74
$Zr_{70}Ni_{10}Pt_{20}$ amorphous	8.80
$Zr_{70}Ni_{10}Pt_{20}$ quasicrystal	9.37
$Zr_{70}Ni_{10}Au_{20}$ amorphous	9.01

of formation of quasicrystals in Zr and noble metal based alloys. Chapters 1 and 2 describe the fundamental thermodynamics and structure of nano-crystalline alloys, and chapters 3–7 and 9 describe the manufacture and structure of a number of different nanocrystalline metallic materials.

Alloy manufacture

Ribbon samples 0.03 mm thick and 1 mm wide have been produced by melt-spinning from $Zr_{70}Ni_{10}Pt_{20}$, $Zr_{80}Pt_{20}$, $Zr_{70}Ni_{10}Au_{20}$ and $Zr_{70}Pd_{30}$ alloys initially prepared by arc melting mixtures of pure metals in an argon atmos-phere. $Zr_{70}Ni_{10}Pt_{20}$, $Zr_{70}Ni_{10}Au_{20}$ and $Zr_{70}Pd_{30}$ ribbons are amorphous and $Zr_{80}Pt_{20}$ ribbons are quasicrystalline in the as-quenched state. Quasi-crystalline $Zr_{70}Ni_{10}Pt_{20}$ and $Zr_{70}Pd_{30}$ ribbons are then prepared by annealing the as-quenched amorphous ribbons at the onset temperature for primary crystallization, 120 s at 800 and 700 K respectively. Alloy densities measured by Archimedes' method are summarized in table 8.1. The alloy manufacture is described in detail elsewhere [9–11].

Anomalous x-ray scattering

The anomalous x-ray scattering measurements were carried out at BL-7C and BL-9C in the Photon Factory of the Institute of Materials Structure Science (IMSS) at the High Energy Accelerator Research Organization (KEK), Tsukuba, Japan. Local atomic structures and radial distribution functions around the noble metals were determined by anomalous x-ray scattering measurements at Pt L_{III}, Au L_{III} and Pd K absorption edges in amorphous $Zr_{70}Ni_{10}Pt_{20}$, $Zr_{80}Pt_{20}$, $Zr_{70}Ni_{10}Au_{20}$ and $Zr_{70}Pd_{30}$ respectively. Local atomic structures and radial distribution functions around the zirconium atoms were determined by anomalous scattering measurements at Zr K absorption edges. The anomalous x-ray scattering intensities were

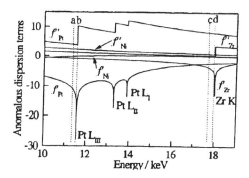

Figure 8.1. Theoretical energy dependence of anomalous dispersion terms for Zr, Ni and Pt near Pt L_{III} and Zr K absorption edges.

measured at 50 and 300 eV below the Zr K and Pd K absorption edges and at 30 and 300 eV below the Pt L_{III} and Au L_{III} absorption edges.

As an example, the variations of anomalous dispersion terms for Zr, Ni and Pt near Zr K and Pt L_{III} absorption edges are shown in figure 8.1. The incident energies used in the present measurements are labelled a–d in figure 8.1. Because of a large change of the real part of the anomalous dispersion term for Pt, f_{Pt}, the observed intensities at a and b show a difference. The energy differential interference function for Pt, $Q\Delta i_{Pt}(Q)$ can then be calculated from this intensity difference using the following equations [15]:

$$Q\Delta i_{Pt}(Q) = \frac{(I(Q)_a - \langle f^2(Q)_a\rangle) - (I(Q)_b - \langle f^2(Q)_b\rangle)}{W(Q)} \quad (8.1)$$

$$W(Q) = \sum_{j=1}^{N} x_j \mathrm{Re}(f_j(Q)_a + f_j(Q)_b) \quad (8.2)$$

$$\langle f^2(Q)\rangle = \sum_{j=1}^{N} x_j f_j^2 \quad (8.3)$$

where I is the coherently scattered intensity, x_j is atomic concentration of element j, f is x-ray atomic scattering factor, N is the total number of elements, Re is the real part of the function, and subscripts a and b indicate incident energies in figure 8.1. The Fourier transformation of $Q\Delta i_{Pt}(Q)$ in equation (8.1) gives the radial distribution function (RDF) for Pt:

$$2\pi r^2 \rho_{Pt}(r) = 2\pi r^2 \rho_0 + \frac{1}{x_{Pt}(f_{Pt}(Q)_a - f_{Pt}(Q)_b)} \int_0^{Q_{max}} Q\Delta i_{Pt}(Q)\sin(Qr)\,\mathrm{d}Q \quad (8.4)$$

$$\rho_{Pt}(r) = \sum_{j=1}^{N} \frac{x_j \mathrm{Re}(f_j(Q)_a + f_j(Q)_b)}{W(Q)}\rho_{Ptj}(r) \quad (8.5)$$

where ρ_{Pt} and ρ_{Ptj} are the number density function around Pt and the partial number density function for Pt-j pairs, respectively. Similarly, the energy differential interference function and radial distribution function for Zr were evaluated from the intensity difference at c and d in figure 8.1.

Coordination numbers and atomic distances around Zr and Pt were evaluated by the method [16, 17] developed for the analysis of the energy differential interference function in equation (8.1). According to Narten and Levy [18], the ordinary interference function is given by

$$Qi(Q) = \sum_{j=1}^{N} \sum_{k=1}^{N} x_j \frac{f_j f_k N_{jk}}{\langle f \rangle^2 r_{jk}} \exp(-b_{jk}Q^2) \sin Qr_{jk}$$
$$+ \sum_{l=1}^{N} \sum_{m=1}^{N} 4\pi\rho_0 x_l x_m \frac{f_l f_m}{\langle f \rangle^2} \exp(-B_{lm}Q^2) \frac{QR_{lm} \cos QR_{lm} - \sin QR_{lm}}{Q^2}$$

$$(8.6)$$

where N_{jk}, r_{jk} and b_{jk} are the coordination number, atomic distance and mean square variation of j–k pairs, and R_{lm} and B_{lm} are the size and variation of the boundary region of l–m pairs. The first term of the right-hand side of equation (8.6) correspond to the discrete Gaussian-like distribution in a near-neighbour region, and the second term corresponds to a continuous distribution with the average number density ρ_0 at a longer distance respectively. Atomic distances and coordination numbers of interest in the near-neighbour region are determined by the least-squares method in equation (8.6) so as to reproduce an experimental interference function. The differential interference function in equation (8.1) is calculated as a difference between the ordinary interference functions computed at the two energies in figure 8.1, just below each absorption edge. Comparing it with the experimental function gives the atomic distances and coordination numbers around any element selected in the anomalous x-ray scattering measurement.

Intensity profiles

Intensity profiles from amorphous $Zr_{70}Ni_{10}Pt_{20}$ and $Zr_{60}Ai_{15}Ni_{25}$ [19] alloys are compared in figure 8.2. Both intensity profiles are basically characteristic of an amorphous phase, i.e. a broad first peak followed by several weaker and broader peaks. The alloy containing Pt, however, is distinct from the alloy without Pt, showing a pronounced prepeak at $16.8 \, nm^{-1}$ and a quite sharp main peak at $25.5 \, nm^{-1}$. A similar intensity pattern is also observed in a quaternary amorphous $Zr_{70}A_{16}Ni_{10}Pt_{14}$ alloy [12]. In general, these features indicate the existence of strong compositional short range order between unlike-atomic pairs in an amorphous alloy [20]. Anomalous x-ray scattering intensity profiles at zirconium and noble metal absorption edges

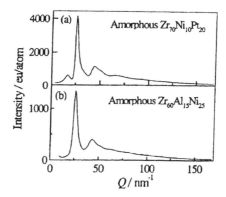

Figure 8.2. Comparison of intensity profiles in (a) amorphous $Zr_{70}Ni_{10}Pt_{20}$ and (b) $Zr_{60}Al_{15}Ni_{25}$ [19] alloys.

in amorphous $Zr_{70}Pd_{30}$, $Zr_{70}Ni_{10}Pt_{20}$ and $Zr_{70}Ni_{10}Au_{20}$ alloys and their energy differential intensity profiles are shown in figures 8.3, 8.4 and 8.5 respectively. Intensity profiles in amorphous $Zr_{70}Pd_{30}$ and $Zr_{70}Ni_{10}Au_{20}$ alloys also show a prepeak and a relatively sharp first peak at 16.8 and 25.5 nm respectively. Similar features are also observed in the intensity profiles from amorphous $Al_{77.5}Mn_{22.5}$, $Al_{56}Si_{30}Mn_{14}$ and $Al_{75}Cu_{15}V_{10}$ alloys [20–22] that are transformed to an icosahedral quasicrystalline phase by heat-treatment. Furthermore, the positions of the prepeaks and the first peaks show no significant change for different noble metals. This implies that the basic local atomic structures are not appreciably affected by the noble metal.

Energy differential intensity profiles of these amorphous alloys obtained at Zr and noble metal absorption edges are also shown in figures 8.3(b) and

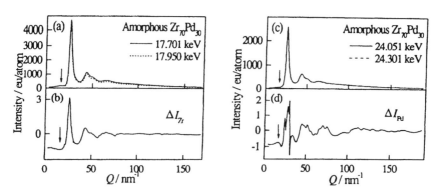

Figure 8.3. Anomalous x-ray scattering and energy differential intensity profiles of an amorphous $Zr_{70}Pd_{30}$ alloy observed at Zr K ((a) and (b)) and Pd K absorption edges ((c) and (d)).

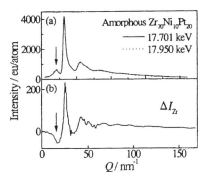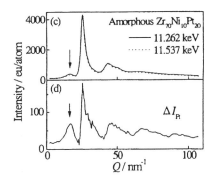

Figure 8.4. Anomalous x-ray scattering and energy differential intensity profiles of an amorphous $Zr_{70}Ni_{10}Pt_{20}$ alloy observed at Zr K ((a) and (b)) and Pt L_{III} absorption edges ((c) and (d)).

(d), 8.4(b) and (d), and 8.5(b) and (d). Structural information on local atomic structures around Zr and noble metals is reflected in them. Although the prepeak indicated with the arrows in figure 8.3(a) disappears in the differential profile at the Zr K absorption edge in (b), it is present in the profile at the Pd K absorption edge of (d). These changes in the energy differential profiles are common in all the three alloys. This suggests that the prepeak is attributed to some short range ordered clusters formed around the noble metals. The intensities of the prepeaks in amorphous $Zr_{70}Ni_{10}Pt_{20}$ and $Zr_{70}Ni_{10}Au_{20}$ alloys are much larger than that in the amorphous $Zr_{70}Pd_{30}$ alloy. This is explained by a difference in x-ray atomic scattering factor between Zr and Au or Pt which is much larger than the difference between Zr and Pd.

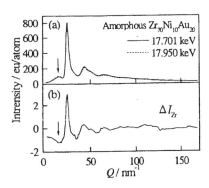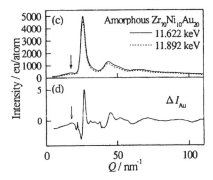

Figure 8.5. Anomalous x-ray scattering and energy differential intensity profiles of an amorphous $Zr_{70}Ni_{10}Pt_{20}$ alloy observed at Zr K ((a) and (b)) and Au L_{III} absorption edges ((c) and (d)).

Coordination numbers and atomic separations

Differential interference functions for Zr and Pt, $Q\Delta i_{Zr}$ and $Q\Delta i_{Pt}$, and ordinary interference functions Qi in the quasicrystalline $Zr_{80}Pt_{20}$ alloy are shown in figure 8.6. The solid and dotted curves in figure 8.6 correspond to the experimental and calculated functions. In figure 8.6, the ordinary interference function including all atomic pair correlations is evaluated from the intensity data observed at c in figure 8.1, i.e. 300 eV below the Zr K absorption edge. Radial distribution functions for Zr and Pt in figure 8.7 were evaluated by Fourier transformation of the functions in figure 8.6. The solid and dotted curves in figure 8.7 correspond to the experimental and calculated functions. First, coordination numbers and atomic distances for Pt-Pt and Pt-Zr pairs are calculated by fitting $Q\Delta i_{Pt}$ in figure 8.6(b). Second, those for Zr-Zr pairs are evaluated from $Q\Delta i_{Zr}$ in figure 8.6(a) using the values for Zr-Pt pairs obtained in $Q\Delta i_{Pt}$. Finally, the function Qi in figure 8.6(c) is calculated from the data of all Pt-Pt, Pt-Zr and Zr-Zr pairs in order to confirm that the present anomalous x-ray scattering analyses provide a correct result. Coordination numbers in nearest-neighbour regions in the other alloys are calculated similarly, and are summarized in table 8.2.

The atomic distances of Zr-Zr and Zr-Ni pairs are almost equal or slightly longer than those calculated from the atomic radii of 0.160 nm for Zr and 0.124 nm for Ni. On the other hand, the atomic distances of Zr and

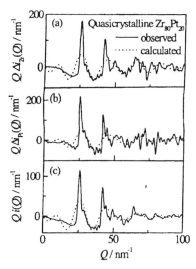

Figure 8.6. Energy differential interference functions for (a) Zr and (b) Pt and (c) ordinary interference functions in a quasicrystalline $Zr_{80}Pt_{20}$. The solid and dotted curves in figure 8.4 correspond to the experimental and calculated ones.

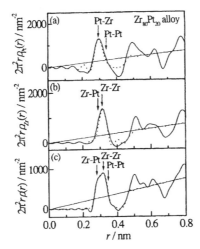

Figure 8.7. Environmental radial distribution functions (RDFs) for (a) Pt and (b) Zr, and (c) ordinary RDFs in a quasicrystalline $Zr_{80}Pt_{20}$. The solid and dotted curves in figure 8.6 correspond to the experimental and calculated ones.

noble metal pairs are much shorter than those evaluated from the atomic radii of Zr and the noble metals. Namely, the experimental distances are 5.7 and 3.7% shorter in the amorphous and quasicrystalline $Zr_{70}Pd_{30}$ alloys, 2.7% shorter in the quasicrystalline $Zr_{80}Pt_{20}$ alloy, 3.3 and 4.0% shorter in the amorphous and quasicrystalline $Zr_{70}Ni_{10}Pt_{20}$ alloys, and 5.6% shorter in the amorphous $Zr_{70}Ni_{10}Au_{20}$ alloy. This indicates that there are strong chemical correlations between Zr and the noble metals as expected from the large negative heats of mixing of Zr and noble metals. This result is also consistent with the presence of the prepeaks in the energy differential profiles for the noble metals.

Total coordination numbers around Zr, $N_T(Zr)$, and the noble metals, $N_T(M)$ in the nearest-neighbour region are calculated from the coordination numbers in table 8.2 and are summarized in table 8.3. There are some interesting results in table 8.3. In the $Zr_{70}Pd_{30}$ alloy, $N_T(Pd)$ in the amorphous state increases to about 12 in the quasicrystalline state while $N_T(Zr)$ does not change in either state. In the ternary $Zr_{70}Ni_{10}Pt_{20}$ alloy, both $N_T(Zr)$ and $N_T(Pt)$ are almost equal to 12 in the amorphous state, and $N_T(Zr)$ increases in the quasicrystalline state while $N_T(Pt)$ shows no change. In the binary quasicrystalline $Zr_{80}Pt_{20}$ alloy, $N_T(Pt)$ is close to 12 and the $N_T(Zr)$ is close to 14. These values resemble those in the ternary quasicrystalline alloy. In the amorphous $Zr_{70}Ni_{10}Au_{20}$ alloy, $N_T(Zr)$ and $N_T(Au)$ are about 14 and 10, respectively. This resembles the values in the other amorphous alloys.

Table 8.2. Coordination numbers and atomic distances of (a) amorphous and quasicrystalline $Zr_{70}Pd_{30}$ alloys, (b) quasicrystalline $Zr_{80}Pt_{20}$ alloy, (c) amorphous and quasicrystalline $Zr_{70}Ni_{10}Pt_{20}$ alloys and (d) amorphous $Zr_{70}Ni_{10}Au_{20}$ alloy.

(a) $Zr_{70}Pd_{30}$ alloys

Amorphous phase			Quasicrystalline phase		
Pairs	r (nm)	N	Pairs	r (nm)	N
Zr-Zr	0.320	11.5	Zr-Zr	0.326	10.3
Zr-Pd	0.280	2.3	Zr-Pd	0.286	3.1
Pd-Pd	0.307	3.6	Pd-Pd	0.304	4.6

(b) Quasicrystalline $Zr_{80}Pt_{20}$ alloy

Pairs	r (nm)	N
Zr-Zr	0.323	11.8
Zr-Pt	0.291	2.4
Pt-Pt	0.328	2.9

(c) $Zr_{70}Ni_{10}P_{20}$ alloys

Amorphous phase			Quasicrystalline phase		
Pairs	r (nm)	N	Pairs	r (nm)	N
Zr-Zr	0.327	8.1	Zr-Zr	0.328	9.7
Zr-Ni	0.327	1.2	Zr-Ni	0.328	1.4
Zr-Pt	0.289	2.8	Zr-Pt	0.287	2.5
Pt-Pt	0.333	1.4	Pt-Pt	0.325	2.5

(d) Amorphous $Zr_{70}Ni_{10}Au_{20}$ alloy

Pairs	r (nm)	N
Zr-Zr	0.331	9.4
Zr-Ni	0.300	2.4
Zr-Au	0.288	2.7
Au-Au	0.305	0.9

Crystallization

In the quenched amorphous $Zr_{70}Ni_{10}Pt_{20}$, crystallization can be described as [9]:

$$\text{amorphous} \longrightarrow \text{icosahedral quasicrystalline} \longrightarrow Zr_2Ni + ZrPt$$

Table 8.3. Total coordination numbers around Zr (N_{Zr}) and around the noble metals (N_M).

	N_{Zr}	N_M
Amorphous $Zr_{70}Pd_{30}$	13.8	9.0
Quasicrystal $Zr_{70}Pd_{30}$	13.4	11.7
Quasicrystal $Zr_{80}Pt_{20}$	14.2	12.4
Amorphous $Zr_{70}Ni_{10}Pt_{20}$	12.1	11.7
Quasicrystal $Zr_{70}Ni_{10}Pt_{20}$	13.6	11.8
Amorphous $Zr_{70}Ni_{10}Au_{20}$	14.5	10.4

The atomic structure of the Zr_2Ni crystal is shown in figure 8.8 [24]. This crystalline structure is face-centred cubic, $Fd3m$ with a large lattice constant of 1.227 nm [25], and contains 64 Zr and 32 Ni atoms. This is sometimes called the 'big-cube' structure, and it consists of icosahedral clusters of Zr and Ni atoms. As an example, two clusters around Zr and Ni are picked up from the unit cell and shown in figure 8.8(b). Consequently, in the $Zr_{70}Ni_{10}Pt_{20}$ system, the quasicrystalline phase decomposes into a ZrNi compound consisting of icosahedral clusters of Zr and Ni, and a ZrPt compound.

Considering (a) the presence of icosahedral clusters around Pt in the amorphous and quasicrystalline states and (b) the structural feature of the crystalline intermetallic compound Zr_2Ni indicates the mechanism of crystallization in the ternary alloy, as follows. In the amorphous state, there are icosahedral clusters around Pt consisting of Zr, Ni and Pt atoms. The quasicrystalline state is also composed of icosahedral clusters similar to those in the amorphous state. These clusters transform to icosahedral

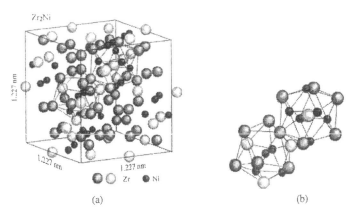

(a)   (b)

Figure 8.8. Atomic structures of the Zr_2Ni crystal [24]. Unit cell and (b) examples of icosahedral clusters around Zr and Ni atoms in the unit cell.

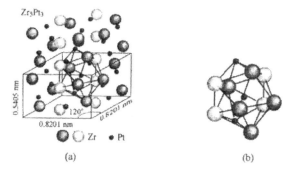

(a) (b)

Figure 8.9. Atomic structures of the Zr_5Pt_3 crystal [26]. (a) Unit cell and (b) polyhedron formed around Pt.

clusters of Zr and Ni atoms in Zr_2Ni by diffusing Pt atoms out of the clusters in the quasicrystalline phase.

On the other hand, in the binary $Zr_{80}Pt_{20}$ system, the quasicrystalline phase decomposes into three crystalline phases [10].

$$\text{amorphous} \longrightarrow \text{icosahedral quasicrystalline} \longrightarrow Zr + ZrPt + Zr_5Pt_3$$

In the binary alloys, the crystalline phases do not exhibit icosahedral clusters like those in Zr_2Ni as shown in figure 8.8(b). In the Zr_5Pt_3 crystal, however, a polyhedron around Pt consisting of nine Zr and two Pt atoms is very similar to an icosahedral cluster. The crystalline structure of Zr_5Pt_3 [26] and the polyhedron are shown in figures 8.9(a) and (b) respectively. Thus, it appears that decomposition of the quasicrystal in the binary system is also explained by the icosahedral clusters around Pt.

The same idea in the present Zr-Pt system might explain crystallization in other alloy systems. For example, crystallization in $Zr_{70}Ni_{10}Au_{20}$ can be described as follows [9]:

$$\text{amorphous} \longrightarrow \text{icosahedral quasicrystalline} \longrightarrow Zr_3Au$$

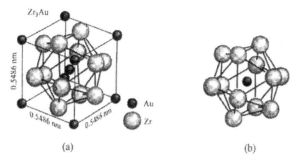

(a) (b)

Figure 8.10. Atomic structures of the Zr_3Au crystal [27]. (a) Unit cell and (b) an icosahedral cluster formed around Au.

The atomic structure of Zr_3Au is shown in figure 8.10 [27]. In Zr_3Au, 12 Zr atoms form an icosahedral cluster around Au atoms that are distributed on a body-centred cubic lattice. The crystalline structure is composed of icosa-hedral clusters around the noble metal. With icosahedral clusters in the $Zr_{70}Ni_{10}Au_{20}$ quasicrystal, similar to Zr-Pt alloys, the crystallization process is straightforward, as shown above. The crystallization temperature in $Zr_{70}Ni_{10}Au_{20}$ is, however, about 30 K higher than that in $Zr_{70}Ni_{10}Pt_{20}$ [9]. Thus, the role of Ni atoms during crystallization process is important to explain both the thermal stability and the atomic structures in the quasi-crystalline phase.

Conclusions

Analyses of the local atomic structures around Pd, Pt and Au and Zr in amorphous and quasicrystalline $Zr_{70}Pd_{30}$, $Zr_{80}Pt_{20}$, $Zr_{70}Ni_{10P}Pt_{20}$ and $Zr_{70}Ni_{10}Au_{20}$ alloys, and examination of related crystalline structures such as cubic Zr_2Ni, Zr_3Au and Zr_5Pt_3 show the presence of strong compositional short range order with clusters around the noble metal atoms in the amorphous state. This is also shown by short atomic separations between Zr and noble metal atom pairs. Formation and decomposition of the quasi-crystalline phase are, therefore, closely related with the presence of icosa-hedral clusters around the noble metal atoms.

References

[1] Inoue A, Zhang T and Masumoto T 1990 *Mater. Trans. JIM* **31** 177

[2] Koster U, Meinhardt J, Roos S and Liebertz H 1996 *Appl. Phys. Lett.* **69** 179

[3] Xing L Q, Eckert J, Löser W and Schultz L 1999 *Appl. Phys. Lett.* **74** 664

[4] Eckert J, Mattem N, Zinkevitch M and Seidel M 1988 *Mater. Trans. JIM* **39** 623

[5] Koster U, Meinhardt J, Roos S and Rudiger A 1996 *Mater. Sci. Forum* **225–227** 311

[6] Chen M W, Zhang T, Inoue A, Sakai A and Sakurai T 1999 *Appl. Phys. Lett.* **75** 1697

[7] Inoue A, Zhang T, Saida J, Matsushita M, Chen M W and Sakurai T 1999 *Mater. Trans. JIM* **40** 1181

[8] Inoue A, Saida J, Matsushita M and Sakurai T 2000 *Mater. Trans. JIM* **41** 362

[9] Saida J, Matsushita M and Li C 2000 *Appl. Phys. Lett.* **76** 3558

[10] Saida J, Matsushita M and Li C 2000 *Appl. Phys. Lett.* **77** 73

[11] Saida J, Matsushita M and Inoue A 2001 *Appl. Phys. Lett.* **79** 412

[12] Matsubara E, Sakurai M, Nakamura T, Imafuku M, Sate S, Saida J and Inoue A 2001 *Scripta Mater.* **44** 2297

[13] Takagi T, Ohkubo T, Hirotsu Y, Murty B S, Hono K and Shindo D 2001 *Appl. Phys. Lett.* **79** 485

[14] Saida J, Matsushita M, Li C and Inoue A 2001 *Phil. Mag. Lett.* **81** 39

[15] Matsubara E and Waseda Y 1994 *Resonant Anomalous X-Ray Scattering Theory and Applications* ed. G Materlik, C J Sparks and K Fischer (Elsevier) pp 345–364

[16] Matsubara E, Sugiyama K, Waseda Y, Ashizuka M and Ishida E 1990 *J. Mater. Sci. Lett.* **9** 14

[17] Waseda Y, Matsubara E, Okuda K, Ornate K, Tohji K, Okuno S N and Inomata K 1992 *J. Phys.: Condens. Matter* **4** 6355

[18] Narten A H and Levy H A 1969 *Science* **160** 447

[19] E Matsubara, T Tamura, Y Waseda, A Inoue, T Zhang and T Masumoto 1992 *Mater. Trans. JIM* **33** 873

[20] H S Chen, D Koskenmaki and C H Chen 1987 *Phys. Rev. B* **35** 3715

[21] Matsubara E, Harada K, Waseda Y, Chen H S, Inoue A and Masumoto T 1988 *J. Mater. Sci.* **23** 753

[22] Matsubara E, Waseda Y, Tsai A P, Inoue A and Masumoto T 1988 *Z. Naturforsch.* **43a** 505

[23] Matsubara E, Waseda Y, Tsai A P, Inoue A and Masumoto T 1990 *J. Mater. Sci.* **25** 2507

[24] Villars P and Calvert L D (eds) 1991 *Pearson's Handbook of Crystallographic Data for Intermetallic* Phases vol 3 (Materials Park, OH: ASM) p 3421

[25] Altounian Z, Batalla E, Strom-Olsen J and Walter J 1987 *J. Appl. Phys.* **61** 149

[26] Villars P and Calvert L D (eds) 1991 *Pearson's Handbook of Crystallographic Data for Intermetallic Phases* vol 4 (Materials Park, OH: ASM) p 5021

[27] Villars P and Calvert L D (eds) 1991 *Pearson's Handbook of Crystallographic Data for Intermetallic Phases* vol 1 (Materials Park, OH: ASM) pp 1355–1356

Chapter 9

Quasicrystalline nanocomposites

Won Tae Kim

Introduction

Quasicrystals have been reported to form in many alloy systems and exhibit characteristic physical and mechanical properties responsible: low electric conductivity, low thermal diffusivity, high hardness, low surface energy and low coefficient of friction [1]. Even though some progress has been achieved in the field of the application of quasicrystals as a non-sticking coating material, full application of these properties into an engineering material is limited due to brittleness. A possible method of solving this inherent brittleness problem is to use composites consisting of quasicrystalline particles embedded in a ductile matrix phase. The mechanical properties of a composite are controlled by the interface between the reinforcing and matrix phases, and property maximization may be obtained by refining the quasicrystalline particles. Indeed homogeneous distribution of nanoscaled quasicrystalline particles in a crystalline or amorphous matrix results in a significant improvement in mechanical properties [2, 3].

The atomic structure of a quasicrystalline phase exhibits no translational order but only rotational order. Due to the structural similarity between quasicrystalline and amorphous phases, interfacial energies between quasicrystalline and crystalline phases (as with amorphous and crystalline phases) are expected to be small [4]. This may contribute to improved thermal stability of a nanoscaled microstructure, because the driving force for particle coarsening is inversely proportional to the particle size and proportional to the interfacial energy.

Several advanced Al alloys, showing excellent mechanical properties, have been developed by distributing nanoscaled quasicrystalline particles in an fcc aluminium matrix [3]. High elevated temperature strength aluminium alloy can be developed in the Al-Fe-Cr-Ti system. The alloy fabricated by hot extrusion of gas atomized powders shows a microstructure consisting of 400 nm sized icosahedral phase particles and 500 nm sized Al grains. The tensile fracture strength is 400–460 MPa at 473 K and 350–360 MPa at 573 K. The high elevated temperature strength is maintained even after annealing for 200 h at 573 K, indicating thermal stability of the microstructure. High strength can also be developed in the Al-Cr-Ce-M alloy system. Rapidly solidified alloy ribbons have a microstructure of 30–50 nm sized icosahedral phase dendrites embedded in an fcc Al phase, with a fracture strength of 1320 MPa and good bending ductility. High ductility can be developed in the Al-Cr-Ce-Co system, with fracture strengths of 500–600 MPa and plastic elongations of 12–30%. The brittleness of the quasicrystalline phase is overcome by dispersion of nanoscaled quasicrystalline particles in a ductile matrix.

This chapter describes the formation of composite microstuctures consisting of quasicrystalline particles embedded in an amorphous or crystalline matrix, and discusses some of the resulting mechanical properties of the composites. For the formation of the quasicrystalline phase, several different processes have been used: glass formation followed by annealing for partial crystallization, mechanical alloying followed by heat treatment, and conventional solidification. Chapters 1 and 2 describe the fundamental thermodynamics and structure of nanocrystalline alloys, and chapters 3–8 describe the manufacture and structure of a number of different nanocrystalline metallic materials. Chapters 13 and 14 describe different kinds of nanocomposite materials.

Partial crystallization

Alloys in the Zr-Cu-Ni-Al system show large glass forming ability. Bulk metallic glasses (BMG) can be fabricated up to 5 mm in diameter by injection casting into a copper mould. The addition of silver to the Zr-Cu-Al-Ni multi-component glass forming alloy increases the precipitation tendency of the quasicrystalline phase [5, 6]. Figures 9.1(a)–(d) show typical bright field transmission electron microscope images of $Zr_{65}Al_{7.5}Cu_{17.5-x}Ni_{10}Ag_x$ alloys with $x = 2.5, 5, 7.5$ and 10 respectively, obtained after annealing treatments for 1800 s at 693 K, and figures 9.1(e) and (f) show micro-electron diffraction patterns taken from the alloy with $x = 2.5$. The electron diffraction patterns (e) and (f) show typical five-fold and two-fold diffraction symmetries respectively, characteristic of the icosahedral quasicrystalline phase. The bright field images show homogeneous distributions of spherical quasicrystals in an amorphous matrix. With increasing silver content, the grain size of the icosahedral phase decreases significantly, as can be seen in figures 9.1(a)–(d), from about 100 to 20 nm as the silver Ag content increases from 2.5 to 10%. An amorphous phase can be seen at the grain boundaries in the alloy with $x = 2.5$, as marked by arrows. Transmission electron microscopy shows that the fraction of icosahedral phase increases with increasing

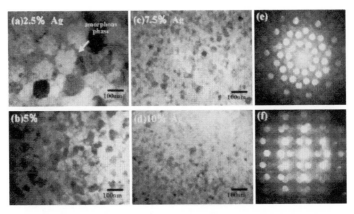

Figure 9.1. (a)–(d) Bright field TEM images of $Zr_{65}Al_{7.5}Cu_{17.5-x}Ni_{10}Ag_x$ (x = 2.5, 5, 7.5 and 10) alloys, annealed for 1800 s at 693 K; (e), (f) micro-electron diffraction patterns after annealing, with x = 2.5.

silver content, indicating that the icosahedral phase composition becomes closer to the matrix composition [6]. The hardness increases linearly with the volume fraction of the quasicrystalline phase. However, the precipitation of a nanoscaled icosahedral phase brings a loss of ductility. The as-quenched amorphous ribbon shows 180° bending ductility, but the annealed ribbon does not show bending ductility and is brittle.

Another example of a composite microstructure consisting of an icosahedral phase distributed in an amorphous matrix can be fabricated by partial crystallization of bulk metallic glass forming $Ti_{40}Zr_{29}Be_{16}Cu_8Ni_7$ alloy. The amorphous alloy exhibits a glass transition temperature of 620 K and crystallizes through two exothermic reactions with onset temperatures of 659 and 750 K. Nanoscaled primary icosahedral phase precipitates from an amorphous matrix and then the remaining amorphous phase crystallizes into the cubic β-Ti(Zr,Ni,Cu) phase [7]. The icosahedral phase transforms into the Laves phase at high temperature by an endothermic reaction, which suggests that the icosahedral phase is stable at lower temperatures. Figure 9.2 shows typical bright-field and dark-field transmission electron microscope images and a selected area electron diffraction pattern (SADP) from $Ti_{40}Zr_{29}Cu_8Ni_7Be_{16}$, heated to 738 K for partial crystallization. The selected area diffraction pattern corresponds to a mixture of quasicrystalline and amorphous phases, in agreement with x-ray diffraction studies [7]. The heat treated alloy shows a homogeneous distribution of icosahedral quasicrystalline phase particles with a size of 5–10 nm. The detailed crystallization behaviour upon annealing at higher temperature is described elsewhere [7].

The mechanical properties of partially quasicrystalline alloys have been measured by compression testing bulk metallic glass specimens, 1 mm in diameter and 2 mm in length, as-machined from as-cast bulk metallic glass

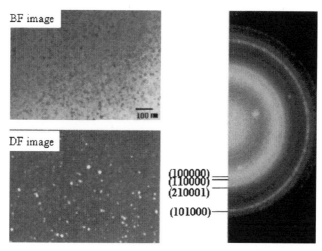

Figure 9.2. Bright-field and dark-field TEM images and corresponding SADPs from melt-spun $Ti_{40}Zr_{29}Cu_8Ni_7Be_{16}$ after heating to 738 K.

samples. The volume fraction of the quasicrystalline precipitate varies with heating up to different temperatures (661, 673, 688 and 715 K). The volume fraction of quasicrystals is difficult to estimate by microstructural analysis since the quasicrystal grains are too small. Instead, the volume fraction can be estimated by comparing the differential scanning calorimeter exothermic heat release of the partially crystallized alloy with the as-cast bulk metallic glasses, although the crystallization heat release rate does depend on the reaction products.

Figure 9.3 shows typical compressive stress–strain curves for as-cast amorphous and partially crystallized $Ti_{40}Zr_{29}Cu_8Ni_7Be_{16}$ alloy rods. The as-cast amorphous alloy shows a yield strength of 1893 MPa with an overall strain of 5.11%. The alloy heated to 661 K has a volume fraction of ~7% and a yield strength of 2038 MPa with an overall strain of 6.15%. The elastic modulus increases because of precipitation of the icosahedral quasicrystalline phase, with a higher hardness and elastic modulus than the amorphous matrix. With further increase in the volume fraction of quasi-crystals the plastic strain decreases, but the compressive fracture stress is still higher than that of the as-cast bulk metallic glass. *In-situ* composites in which nanosized quasicrystals are isolated and homogeneously distributed in an amorphous matrix exhibit a simultaneous improvement in strength and ductility.

Plastic deformation of bulk metallic glasses takes place through shear deformation. Unfortunately, the shear deformation takes place in narrow localized bands due to the absence of strain hardening. The simultaneous increase of strength and compressive strain by partial crystallization of an

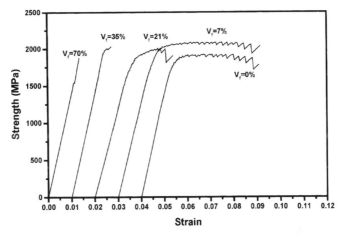

Figure 9.3. Typical compressive stress–strain curves at room temperature for as-cast amorphous and partially crystallized $Ti_{40}Zr_{29}Cu_8Ni_7Be_{16}$ alloy rods.

amorphous phase indicates that the number of shear bands increases. The nanosized quasicrystalline particles can act either as nucleation sites for shear band formation or as barriers to shear band propagation, leading to the increase of yield strength and fracture strength as previously reported [8, 9]. The nature of the interface is one of the factors in determining the mechanical properties of the composite alloys. High interfacial strength can prevent shear band propagation effectively without crack initiation, and can provide a source generating multiple shear bands. For many amorphous alloys, both structural relaxation and partial crystallization cause embrittlement compared with the as-cast fully amorphous material. However, Ti-Zr-Cu-Ni-Be shows simultaneous increases of strength and plastic strain after heating to 661 K, as shown in figure 9.3. Similar improvement of mechanical properties by precipitation of the quasicrystalline phase has also been reported in Zr-based bulk metallic glass forming alloy systems [10].

Mechanical alloying

Another way to form nanoscaled quasicrystalline particles is by mechanical alloying. There are several reports showing icosahedral phase formation during annealing after mechanical alloying Al-based alloy systems [11, 12]. Figure 9.4(a) shows a typical bright-field transmission electron microscope image taken from a $Ni_{41.5}Zr_{41.5}Ni_{17}$ bulk sample prepared by mechanical alloying followed by hot consolidation and annealing for 17 h at 807 K. The bulk material exhibits an equiaxed grain structure, with grains of 100–150 nm in diameter. Spherical particles marked by arrows in figure 9.4(a) are

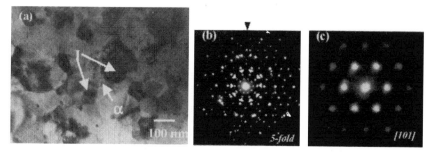

Figure 9.4. Bright field TEM image of bulk $Ti_{40}Zr_{29}Cu_8Ni_7Be_{16}$ after annealing at 530 °C for 15 h, showing icosahedral and α phases.

confirmed to be icosahedral phase by transmission electron microscope analysis as shown in figure 9.4(b), exhibiting the five-fold rotational symmetry of the icosahedral phase. The equiaxed grains are confirmed to be α-solid solution with hexagonal structure as shown by the [110] micro-electron diffraction pattern in figure 9.4(c). Annealing produces a fine-scale composite microstructure consisting of icosahedral phase particles distributed homogeneously in a hexagonal α-solid solution matrix. The microhardness of the bulk material is about 6.2 GPa and cracks are not observed during indentation under loads lower than 10 N [13]. The two-phase microstructure that has a ductile solid solution matrix is technologically important. Through careful control of powder composition and thermal processing parameters, the $(I + \alpha)$ two-phase microstructure can be manufactured.

Conventional solidification

Nanoscaled quasicrystalline particles can be formed during rapid solidification, crystallization of an amorphous phase or mechanical alloying, but the materials need to be consolidated into a bulk form for engineering applications. Bulk composite material consisting of a stable quasicrystalline phase and a ductile matrix can be fabricated by conventional casting, even though the advantages of a fine-scale microstructure are lost. *In-situ* composite material consisting of icosahedral phase distributed in a ductile α-magnesium matrix can be fabricated by conventional casting of Mg-Zn-Y alloys [14, 15].

Figures 9.5(a) and (b) show a typical bright field transmission electron microscope image of a $Mg_{86}Zn_{12}Y_2$ alloy and a corresponding selected area diffraction pattern (SADP) obtained from the dark phase in the eutectic structure. The bright-field image clearly shows dendritically grown primary α-magnesium phase and a eutectic structure in the interdendritic regions. The selected area diffraction pattern shows an aperiodic distribution of diffraction spots, which is characteristic of the quasicrystalline structure.

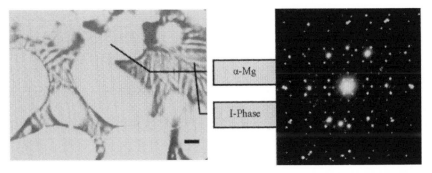

Figure 9.5. Typical bright field TEM image and SADP from the dark phase in the eutectic structure of $Mg_{86}Zn_{12}Y_2$.

The selected area diffraction pattern corresponds to a two-fold axis diffraction pattern from the icosahedral phase.

Figure 9.6(a) shows a scanning electron microscopy (SEM) image of an as-cast $Mg_{95}Zn_{4.3}Y_{0.7}$ alloy. Conventional thermomechanical processing by hot-rolling and annealing are used to distribute the icosahedral phase into a large number of small particles in the α-magnesium matrix. The alloy is hot-rolled from 12 to 1.7 mm final thickness with a reduction per pass of $\sim$15% to a total reduction of 86%. Equiaxed 14 μm-sized grains form by recrystallization during hot rolling, and the eutectic icosahedral solidification structure is destroyed, providing the microscale particles shown in figure 9.6(b) of the scanning electron micrograph after hot-rolling. The broken particle/particle interfaces are found to be completely healed by the thermally activated processes.

The hot-rolled Mg alloy sheets have been annealed at 673 K for 0.5 h in an air circulating furnace, and uniaxial tension tests conducted at room

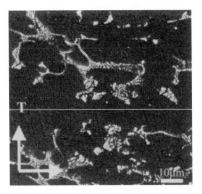

Figure 9.6. SEM images of (a) as-cast and (b) hot-rolled $Mg_{95}Zn_{4.3}Y_{0.7}$ (L and T are longitudinal and transverse directions).

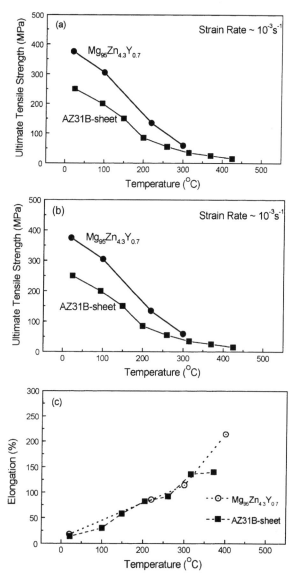

Figure 9.7. Effect of temperature on (a) UTS, (b) yield stress ($\varepsilon = 0.2\%$) and (b) elongation for $Mg_{95}Zn_{4.3}Y_{0.7}$ and AZ31B at a strain rate of $10^{-3}\,s^{-1}$.

temperature and elevated temperature up to 673 K under a constant strain rate of 10^{-3} and $10^{-4}\,s^{-1}$. The resulting tensile and yield stresses and elongations are plotted as a function of test temperature in figures 9.7(a)–(c), respectively. For comparison, corresponding values of AZ31B, which is one of the most commonly used wrought Mg alloys, are also given. In

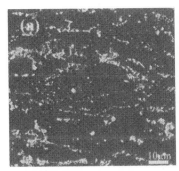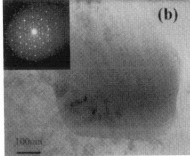

Figure 9.8. (a) SEM and (b) bright-field TEM images of an icosahedral particle after deformation to a strain of 1.0 at a strain rate of $10^{-4}\,\mathrm{s}^{-1}$ and test temperature of 673 K.

the low-temperature regime, typically up to 473 K, the $Mg_{95}Zn_{4.3}Y_{0.7}$ alloy exhibits much higher strength than AZ31B, without decreasing the elongation. With increasing temperature, tensile and flow strengths both decrease, mainly due to dislocation motion on non-basal planes in the α-magnesium matrix and the contribution of grain boundary sliding to the total deformation. However, the flow strength of the alloy is still high when compared with AZ31B. In general, alloys with larger amounts of second phase show lower levels of elongation in the metal matrix composites, since the frequency of crack initiation sites increases.

Figures 9.8(a) and (b) show, respectively, scanning and bright-field transmission electron microscope images after deformation to a strain of around 1.0 at a strain rate of $10^{-4}\,\mathrm{s}^{-1}$ and a test temperature of 673 K. The flow stress of the alloy is constant at $\sim$10 MPa over the strain range and it fails by diffuse necking. Particles initially distributed in or near eutectic regions before testing (see figure 9.7) move away from each other during deformation, being distributed more randomly in the test specimen. During testing at 673 K, there is no significant growth of the quasicrystalline particles. No debonding or nanoscale defects are observed at the interface between the particles and the matrix.

It is believed that all the improved mechanical properties observed in this alloy are mainly due to the stable icosahedral phase particles. The icosahedral phase exhibits more isotropic characteristic than corresponding crystalline phases because of its high symmetry. This provides reasonably stable bonding with the matrix, and low strain energy at the matrix/particle interface and near the icosahedral phase particle. In general, the matrix near microscale intermetallic particles is highly stressed due to the mismatched lattice constants between the matrix and the particle. However, due to the quasiperiodic lattice structure of icosahedral phase, the mismatched strain may be compromised by the icosahedral phase particle, decreasing the stress concentration in the matrix near the icosahedral phase particle.

A low quasicrystal interfacial energy reduces the driving force for growth of the icosahedral particles. Moreover, low diffusivity of yttrium in magnesium is expected given the large difference in atomic sizes, leading to a stable quasicrystal particle size of the particles. These factors are probably responsible for negligible coarsening and a stable quasicrystal particle size during high temperature deformation, as shown in figure 9.8(a).

Summary

Composite materials consisting of quasicrystalline particles embedded in an amorphous or ductile crystalline matrix can be fabricated by various different routes: partial crystallization of amorphous phase; rapid solidification and hot consolidation; mechanical alloying and hot consolidation; and conventional casting followed by thermomechanical treatment. The composites materials show improved mechanical properties at room and elevated temperatures. The interfaces between the matrix and the quasicrystalline particles are very stable and this is effective in improving the mechanical properties.

References

[1] Janot C 1994 *Quasicrystals: a Primer* 2nd edition (Oxford: Clarendon Press) pp 1–52
[2] Xing L Q, Eckert J, Löser W and Schultz L 1999 *Appl. Phys. Lett.* **74** 664
[3] Inoue A 1998 *Prog. Mater. Sci.* **43** 365
[4] Dubois J M, Plaindoux P, Belin-Ferre E, Tamura N and Sordelet D J 1997 *Proc. 6th Int. Conf. on Quasicrystals* ed. Fujiwara and Takeuchi (Singapore: World Scientific)
[5] Inoue A, Zhang T and Masumoto T 1990 *Mater. Trans. JIM* **31** 177
[6] Lee J K, Choi G, Kim W T and Kim D H 2000 *Appl. Phys. Lett.* **77** 978
[7] Kim Y C, Kim W T and Kim D H 2002 *Ann. Chim. Sci. Mat.* **27** 11
[8] Eckert J, Reger-Leonhard A, Weiß B and Heilmaier M 2001 *Mater. Sci. Eng.* A **301** 1
[9] Choi-Yim H, Busch R, Köster U and Johnson W L 1999 *Acta Mater.* **47** 2455
[10] Inoue A, Zhang T, Chen M W and Sakurai T 2000 *J. Mater. Res.* **15** 2195
[11] Eckert J and Schultz L 1991 *Mat. Sci. Eng* A **133** 393
[12] Kim K B, Kim S H, Kim W T, Kim D H 2001 *Mat. Sci. Eng.* A **304–306** 822
[13] Yi S, Kim K B, Kim W T and Kim D H 2001 *Scripta Mater.* **44** 1757
[14] Bae D H, Kim S H, Kim W T and Kim D H 2001 *Mat. Trans. JIM* **42** 2144
[15] Bae D H, Kim S H, Kim D H and Kim W T 2002 *Acta Mater.* **50** 2343

SECTION 2

NOVEL NANOMATERIALS

Nanocrystalline materials can be manufactured in a wide variety of different forms: particulate, layered and bulk materials, in metallic, ceramic, polymer, biological and composite materials. This section describes some of the more novel cases of nanocrystalline materials. Chapter 10 discusses nano-optoelectronic materials; chapters 11 and 12 discuss steels manufactured by heavy plastic deformation; and chapters 13 and 14 discuss respectively metal–ceramic and ceramic–ceramic nanocomposite materials.

Chapter 10

Nano-optoelectronics

Peter Dobson

Introduction

Nanoparticles have many potential applications in a wide range of different fields, but their application to making optical materials is one of the oldest, with examples that have been around for centuries. Small nanometre sized particles of gold and semiconductors such as CdSe have been used to colour glass for over a thousand years, and they have had widespread use in ancient pottery glazes. It is very likely that in the next decade the newer methods of making nanoparticles in a pure and well controlled size and shape will be applied to the creation of some important optoelectronic devices. Most of the basic science concerning nanoparticles is now established, and it is possible to foresee them being used to make materials with a well-defined refractive index, electro-optic modulators, low threshold lasers, new computing elements, luminescent screens and bio-labels amongst other things.

In this chapter, the basic physical properties of nanoparticles of metals and semiconductors will be reviewed, and an outline will be given of the ways in which they can be made and incorporated into composite materials. The application of these new materials will depend on enabling technologies to realize some of the device structures, and some of the main issues will be identified.

Background

The recent excitement and hype concerning the possibilities offered by nanotechnology tends to overlook the fact that there are many examples around that have exploited nanoparticles for centuries. Many of the old beautiful stained glass windows in the Oxford Colleges contain either gold or cadmium sulpho-selenide particles. Glazes in pottery, particularly the famous *Mayan Blue* are believed to owe their properties to a dispersion of nanoparticles. Other articles of antiquity such as Venetian glass and the famous Satsuma glass owe their delicate and unique colours to nanoparticle dispersions of gold and copper respectively [1–3]. As we examine more closely many living objects we find that a high degree or order and organization at the nanometre level exists in nature, for example the natural iridescence of sea shells and the colours on the wings of moths and butterflies are highly ordered sub-micron structures [4–6]. The huge interest that has developed world-wide in nanotechnology is opening the eyes of scientists and technologists to possibilities of engineering a material at the nano level to emphasize particular functional properties. This chapter will concentrate on the optical and opto-electronic aspects.

In order to give the chapter some structure it is probably best to divide the subject up and first discuss what optical effects change with the size of a particle. This will inevitably lead to another subdivision, namely, between the behaviour of metals and semiconductors. Metals do have interesting optical properties, especially when they are in the form of small particles, displaying strong optical absorption corresponding to surface plasmon resonances. Semiconductors, on the other hand, show optical properties that are determined mainly by the quantum size effect and a size-tuned effective energy gap. So the underlying physical effects that are significant are different for metals and semiconductors. The size of semiconductors and other dielectric materials manifests itself at several hundred nanometres too, in the form of optical resonators.

Where size matters

If we examine length scales in condensed matter, we encounter parameters such as the carrier mean free path, the exciton diameter, the effective Bohr radius and the effective wavelength of electromagnetic excitations in the medium. In large chunks of material, all of these length scales are very small as compared with the dimensions of the sample or indeed often of the crystal dimensions. Their effect on the material properties is therefore minimal. This situation changes considerably when the material dimension becomes the same size as these parameters and smaller.

Carrier mean free path is a concept that we use to describe the collision of electrons or holes in a conduction process. The collisions are with the lattice (phonons) or with defects or impurities. In the case of most metals at room temperature, the electron mean free path is ~10–20 nm. In the case of electrical conductivity, this means that the metal film thickness starts to have an effect on reducing the expected conductivity when it is of this size. For optical effects, in metal layers that are composed of nano-particles, electron scattering by the boundaries of the particle starts to become significant at these sizes. Later, when we discuss metals in more detail we will see what effect this has on the optical properties.

Semiconductors can show some very marked size effects. The simple 'particle in a box' model that is used in introductory courses on quantum mechanics can be used here to good effect. This model predicts that, as the size is reduced, the lowest energy electron and hole states are not at the conduction and valence band edges respectively, but they are at energies above and below these band edges. This simple model is really taking into account that the electron states of the semiconductor nanoparticle are described by the same states as for a bulk piece of that material, *perturbed* by the effects of its boundary. This is known as the *envelope approximation* [7], and it provides a convenient way of calculating the size effect on the electronic and optical properties of nanoparticles. It is important here to note that essentially the size of the particles is such that electron (or hole) waves resonate within the particle. If electrons and holes are indeed confined to such a 'box', there is an additional effect that must be taken into account, that is, an exciton will be formed because of the mutual Coulomb attraction between the electron and hole. This will reduce the energy of the system. These effects can be summarized [8] by the equation

$$E_{\text{eff}} = E_{\text{g}} + \frac{h^2}{8r^2}\left(\frac{1}{m_{\text{e}}} + \frac{1}{m_{\text{h}}}\right) - \frac{1.8e^2}{\varepsilon r} + \text{small terms.} \qquad (10.1)$$

Here, the particle radius is r, E_{eff} is the effective band gap, E_{g} is the bulk material band gap, m_{e} and m_{h} are the effective masses of the electrons and holes, h is Planck's constant, e is the electronic charge and ε is the permittivity of the semiconductor.

Awareness of these effects followed the work of Brus [8] and Efros [9]. Subsequent work [10] has all confirmed that virtually every semiconductor system follows the trends predicted here, i.e. as the particle size is reduced the effective optical energy gap increases. These small semiconductor particles are often referred to as 'quantum dots'. The implications of this size dependence will be examined more closely in the section dealing with semiconductors.

There is another size regime that is very important, and that is when the size of the particle becomes such that light waves can resonate within it. This size regime actually has a very familiar manifestation in everyday life. Emulsions that have a 'white' appearance owe this to the resonant scattering of visible light within the emulsion particles. Thus milk containing an emulsion of small droplets of fat within a water-based medium appears white. Paint designers make the scattering size of the titanium dioxide or similar pigments of a size to enhance the scattering power and hence 'covering power' of the paint. More recently, there are examples of this effect being applied to develop new concepts of laser or light emitters. Thus, for example, an aerosol droplet containing a dye can show laser action at a very low optical pump threshold [11]. Lawandy and co-workers [12, 13] have taken this idea further in suggesting that it should be possible to make low threshold lasers in the form of composite materials that combine either dyes or quantum dots with larger optical resonator particles.

Metal nanoparticles

Metals are not the first materials that spring to mind in optics, except for their use in mirrors. Even here, there are some very interesting features. All materials are best described by the frequency dependence of the permittivity. For metals in the visible region, both the real and the imaginary parts of the permittivity are significant and of the same order of magnitude [14, 15]. This translates to give a high reflectivity R via the relationships

$$R = \frac{(n-1)^2 + \kappa^2}{(n+1)^2 + \kappa^2} \tag{10.2}$$

$$n = \left\{ \frac{(\varepsilon_1^2 + \varepsilon_2^2)^{1/2} + \varepsilon_1}{2} \right\}^{1/2} \tag{10.3}$$

$$\kappa = \left\{ \frac{(\varepsilon_1^2 + \varepsilon_2^2)^{1/2} - \varepsilon_1}{2} \right\}^{1/2}. \tag{10.4}$$

In these equations, n and κ are the real and imaginary part of the refractive index respectively.

In order to maximize the reflectivity, a prerequisite is that the real part of the permittivity should be high, and the imaginary part should not be too large. This is because the imaginary component is associated with the absorption of electromagnetic waves, and energy is lost from the beam, eventually being converted into heat. The shiny silvery metals such as silver and aluminium are obviously good 'visible' mirrors. This is especially because of their permittivity behaviour. But even here there are large differences between these two very similar metals. Silver has a lower optical absorption in the near infrared and visible than aluminium, because of its lower values of ε_2 and consequently it has a slightly higher reflectance. However, silver has a large absorptive feature at around 3.8–4.2 eV due to strong optical transitions from the d-electron states to states above the Fermi level. These *oscillator-like* transitions produce a peak in the $\varepsilon_2(\omega)$ function and consequently cause a dramatic reduction in the reflectivity. However, more dramatic still is the addition of this embedded oscillator-like behaviour to the real part of the permittivity. As long as $|\varepsilon_1| \gg 1$ the reflectivity will be high, but if $\varepsilon_1 = 0$ the reflectivity will drop to zero. This occurs at the well known plasmon frequency ω where

$$\omega = \left(\frac{ne^2}{m\varepsilon_0} \right)^{1/2} \tag{10.5}$$

where n is the electron density, e is the electronic charge, m is the free electron mass and ε_0 is the permittivity of free space.

For the case of aluminum, the plasmon frequency is easy to estimate, since it behaves like a free-electron metal and all three valence electrons are 'free'. This gives an energy for the plasmon of 15.6 eV which is far in the ultraviolet, thus making aluminum the metal of choice for optical measurements from the visible to the far ultraviolet. Silver, on the other hand, because of the effects of the d-electron oscillator-like transitions in the near ultraviolet at 290–330 nm, has its 'free electron' behaviour greatly perturbed. The plasmon condition is now defined by $\varepsilon_1 = 0$ and the reflectivity drops dramatically at wavelengths around 330 nm. Silver, then is a good mirror for the infrared and the visible, but its performance in the violet end of the spectrum is greatly diminished. Similar remarks would apply to gold, and for this metal the plasmon effects are shifted farther into the visible region, giving good reflectivity for red and yellow, but poor for green and blue. This is the mechanism primarily responsible for its colour.

The plasmon behaviour of materials is therefore a strong determining factor in their optical behaviour and appearance. We should therefore examine more closely what happens if we make a material in the form of nanoparticles. This brings to the fore the importance of the oscillation in charge density at the surface of a material, that is *surface plasmons*. The condition for exciting these for a *flat surface* bounded by a dielectric of permittivity ε_d is

$$\varepsilon_1 = -\varepsilon_d. \tag{10.6}$$

For a small sphere the condition is

$$\varepsilon_1 = -2\varepsilon_d. \tag{10.7}$$

Larger spheres can have higher order surface plasmon modes

$$\varepsilon_1 = -\varepsilon_d(m+1)/m \tag{10.8}$$

where m is an integer with the lowest mode corresponding to a dipole excitation at $m = 1$. The effect of the surface plasmon absorption is to introduce a strong absorption in the blue/green for gold, thus giving the ruby red colour of colloidal dispersions of gold in glass and in common solvents. Silver colloids are not as attractive since their plasmon resonance lies farther towards the violet; hence a mixture of colours is transmitted, resulting in a brownish appearance.

To a first approximation, these higher order modes will not appear until the particles have a circumference large enough to support integral numbers of wavelengths in the surrounding medium. Actually this is an oversimplification and in practice it is necessary to use the full Mie theory of scattering to estimate the optical absorption for larger spheres [15]. The surface plasmon operates in the lowest mode for particles less than 20 nm for the case of gold and silver [16]. Larger particles have contributions from quadrupole and higher order transitions and this tends to shift the mean value of the plasmon absorption to longer wavelengths (see [16] and references therein). Another factor, however, is the inevitable aggregation of metallic particles in suspensions or the formation of composites. When metal particles are close together, there will be interactions between the surface plasmon dipoles, and there may be a physical joining up of the particles themselves. These effects require further modification to the scattering treatment [17].

The effects of size in surface plasmon scattering are fairly subtle and give two distinct effects. The effect of extra surface plasmon modes giving rise to a 'red shift' is described above. For particles of less than around 20 nm, the electron mean free path becomes important and this contributes to the broadening of the plasmon resonance. Very small metal colloidal particles (1.5–2 nm) exhibit very broad absorptions.

The sensitivity of the surface plasmon resonance condition to the dielectric permittivity of the surroundings is the basis of a large number of biosensing applications. This was first recognized by Giaever [18] and has been extended to commercial instruments such as the Biacore^{TM} SPR analyser [19]. The basis of operation of such detectors is to have a film of metal such as silver or gold, in the form of discrete particles. These particles are then coated with a receptor molecule that will bind selectively to a specified target molecule. When the binding occurs, the local dielectric permittivity ε_d is changed, hence the condition of plasmon resonance is changed and this can be detected with a suitable spectroscopic instrument, or even in some instances by a marked change in the colour of the film. Giaever's original experiments were performed using indium metal 'island' films; but nowadays, recognizing the superiority of silver and gold, these are the materials of choice. This is because the variation of ε_1 for these metals with wavelength is very marked in the visible part of the spectrum. There are no reports of other materials being used in this application, but possible candidates would be titanium nitride and some of the brightly coloured gold alloys such as Al-Au. The application of the effect will go far beyond that of today's Biacore^{TM} instrument, and we can foresee arrays of sensors being developed that will have the capability of screening blood or serum for a large number of antibodies. There are also strong indications that surface plasmon resonance technology will be applied to DNA sequencing [20]. It has been suggested that an array of colloidal gold nanoparticles can be addressed for DNA hybridization with a sensitivity comparable with fluorescent techniques [21].

The plasmon resonance effect could be more widely used in optoelectronics in future. It was once suggested for optical storage media [22]. Another possibility is to mimic its original application in decorative glass and use thin metal colloid-containing layers of glass or polymer as colour filters. This has not so far been applicable largely because the optical density is not high enough. Where the surface plasmon resonance effect is most likely to be exploited is in the ability of small metal particles to act as local magnifiers of the electromagnetic field, i.e. the lightning rod effect. Examples of this have been discussed by Link and El-Sayed [16].

Semiconductor nanoparticles

The properties of semiconductor nanoparticles are mainly governed by the quantum size effect discussed above. That is, the electronic states depend on the size; and decreasing the particle size generally increases the effective energy gap. This is not the only factor that occurs. The electronic states also depend strongly on the shape of the particle. This is complicated, and few examples have been published that correlate the shape and the electronic states. However, it is natural to expect this since essentially the shape will determine the electron wave resonances within the particle. Another very important consequence of the small size and dimensionality of semiconductor particles is that the density of electron states becomes more 'molecule-like' and very high compared with bulk three-dimensional material [23]. Put another way this is equivalent to saying that the *oscillator strength* of a composite material made from semiconductor nanoparticles will be high provided that the particles are embedded in a medium that maintains the confinement of electrons and holes within the particle. This should have important implications for the design of strongly optical absorbing and emitting materials.

There are several examples of direct current electroluminescent devices that have been made using quantum dots. The first of these used the well known conducting and light emitting PPV polymers as a matrix in which CdSe particles were incorporated [24]. The results from these showed a mixture of light emission from the PPV and the quantum dots. Later work by Salata *et al* [25] using CdS embedded in polyvinylcarbazole showed distinct emission from the quantum dots. This type of device is difficult to make and reproduce but, as our understanding of conducting polymers grows and as the problems of incorporation of nanoparticles into matrices made from polymers improves, we should see advances. The huge merit of the approach is that a 'paint' or 'ink' can be made that can be easily applied by cheap spin- or dip-coating techniques.

Two factors that do not receive the attention that is due to them are the effects of doping and of applied electric fields. These are discussed below.

Doping effects

If electrons are doped into a quantum dot, they will occupy the lowest energy states above the original conduction band. This should alter the excitonic behaviour since the electron will help screen against the formation of a bound electron–hole pair produced by incoming photons. It will also have a marked effect on the infrared absorption in that electrons in these ground states are available to be excited to higher quantized states within the quantum dot. Recently there has been good evidence for both of these effects occurring [33]. Doping of the quantum dots themselves is an interesting question. They could be doped by the incorporation of the appropriate impurity into the particles and, given the colloidal methods of synthesis, this could be accidental, due to impure chemicals. A quantum dot of 5 nm size only contains around 1500 molecular units so it might be very easy to envisage the incorporation of impurities. The biggest difficulty may be to guarantee that each quantum dot only has, say, one impurity atom in it! On the other hand, the impurity in question need not actually be incorporated into the quantum dot. It could be close to the particle, and the electron could tunnel from the impurity into the quantum dot. This we term 'proximity doping' and it is similar in principle to the doping in organic polymer materials [34].

As the use of such quantum dots develops, more attention will have to be paid to this issue. Currently most studies have concentrated on the manufacture of colloidal suspensions of quantum dots and their later incorporation into glass or polymer layers. Hardly any account was made of impurities in the particles or their surroundings. One exception to this was in the work of Salata *et al* [25] in which, in order to make a diode light emitter from quantum dots of CdS, great care was taken using dialysis to remove as much alkali impurity as possible. This was found to be a prerequisite to obtain successful light emission from the devices.

Electric field effects

Surprisingly, apart from some excellent work a few years ago [26], there has been little interest in looking at the quantum Stark effect (QSE) in semiconductor particles apart from a definitive study by Bawendi's group [27]. This effect results from the change of shape of the potential 'box' that confines electrons and holes when an electric field is applied. The electric field causes the potential well to change from a rectangular shape to a trapezoidal shape. This lowers the confined particle energies and red-shifts the optical absorption. Use was made of this effect to design many types of electro-optic modulator using epitaxial quantum well layers [28, 29]. Among the more systematic attempts to study the effect in quantum dots were those of Cotter *et al* [26], who looked at electric field effects in specially annealed Schott filter glass that contains quantum dots of Cd(Se,S). One feature of the quantum Stark effect is that the well size needs to be not too small, but of a size where the electric field perturbation changes the confined carrier wavefunctions. Experiments on quantum epitaxial layers showed that the effects were most marked for sizes in the range of 10–25 nm and this is also likely to be the case for particles distributed in a matrix. It is questionable if any particular advantage comes from making an electro-optic device from quantum dots. They could offer great flexibility in manufacture if they can be easily incorporated into a polymer solution and applied by spin- or dip-coating.

Another aspect of composite materials incorporating quantum dots is the electrical conduction process when an electric field is applied across the material [30]. The dots will behave like deep traps for both electrons and holes if the energy levels are appropriately aligned. The electrical conductivity of such a composite will therefore be low and it will depend very much on the average separation of the dots. It should be possible to 'design' a specific variation of conductivity with temperature by playing off the possible inter-dot tunnelling against thermionic emission via the LUMO/HOMO of the surrounding matrix. Such materials should also be photoconductors, although results in this area are difficult to reproduce and are inconclusive.

Passivated surfaces and core/shell structures

Many workers have recognized that quantum dots need some surface protection. Much of the work referred to above, where the quantum dots have been used to make light emitters, used a surface layer whose active role was to eliminate surface electron states. The best illustration of this is probably the work of Salata *et al* [25], who showed that for 'bare' particles of CdS the luminescence was always in the red part of the spectrum, whilst the absorption showed strong excitonic effects in the blue. This is because the radiative recombination occurs via surface states. It was found that surface treatments by the use of thiol layers, hexametaphosphate or a hydrated oxide of cadmium could eliminate the red emission and lead to bright blue photoluminescence. There is now a widespread adoption of various surface capping protocols. The most efficacious of these is the use of tri-octyl-phosphine oxide (TOPO) developed by Bawendi and co-workers [31]. In this method, the colloidal formation is actually conducted in a molten solvent of tri-octyl-phosphine. The method seems to have wide applicability but it does leave an (often unknown) thickness of TOPO on the surface. This could be a problem for carrier transfer in and out of the nanoparticle.

A slightly more controlled methodology involves the growth of a shell of a wider band-gap semiconductor around the outside of the core particle [32]. This too results in a large increase in the luminescence from the core quantum dot. In principle this could lead to a method for making quantum dot composites if the capped quantum dots can be formed into layers and sintered together such that the cores do not touch each other.

One of the most likely ways to make such core/shell particles is to make use of self-assembled polyelectrolytes [35]. These materials self-assemble around a charged surface, leaving a shell of the opposite charge state to the original particle. It is then possible to self-assemble another polyelectrolyte of opposite charge on the first, building up a bi-layer. This process can be continued almost indefinitely. The method is especially appropriate for application to colloidal dispersions, and it is being used for many materials combinations. In our laboratory we have used it to adjust the surface charge of one particle so that other particles of opposite electric charge self-assemble around the original core. Many novel composite particles are now being made, such as luminescent oxide core plus magnetic shell, magnetic core plus gold shell, etc. [36].

Other optoelectronic nanoparticles

It is impossible in such a short review to do adequate justice to all of the nanoparticles that are finding important applications in optoelectronics. A few more are mentioned here for completeness.

Nanoparticle phosphors

These can be made from many of the conventional phosphor compounds, both oxides and sulphides. The advantage of making them in nanoparticle form is that these are generally perfect small crystals free of the defects that exist in phosphors made by the traditional calcining and grinding routes. It has been shown that such particles have big advantages over their bulk counterparts for use at low energies [37, 38].

Nanoparticle layers

It is possible to deposit layers of nanoparticles by a range of techniques to form thin highly uniform layers. These can be used in solar cells [39] and can form the basis for making transparent conducting coatings. There are also examples of their use for making tough scratch-resistant layers and multiple layers for antireflection purposes [40].

Nanoparticle composites

It is possible to design optical materials with fairly well-defined optical properties. The guidelines for doing this were set out about 100 years ago by Otto Weiner, and the principles were recently re-defined by Aspnes [41, 42]. The basis is to use complex permittivity space to account for the volume proportion of the nanoparticles and host matrix, adding Mie scattering to the predicted properties. This has been tested for silver particles in a variety of matrices with some success [43].

Summary

Nanoparticles are going to form the basis for many new materials in the opto-electronics field in the next decade. Most of the physics of their behaviour is now understood, at least at the semi-quantitative level. This will enable the application of nanotechnology to be a solution provider to this important field.

References

[1] Jose Yacaman M, Rendon L, Arenas J and Puche M C S 1996 *Science* **273** 223

[2] Kerker M J. 1985 *Colloid Interface Sci.* **105** 297

[3] Nikai I, Numako C, Hosono H and Yamasaki K 1999 *J. Am. Ceram. Soc.* **82** 689

[4] Gale M 1989 *Physics World* October p 24

[5] Vukusic P, Sambles J R and Lawrence C R 2000 *Nature* **404** 457

[6] Srinivasarao M 1999 *Chem. Rev.* **99** 1935

[7] Weisbuch C 1987 *Semiconductors and Semimetals* **24** 1

[8] Brus L E 1984 *J. Chem. Phys.* **80** 4403

[9] Efros A L and Efros A L 1982 *Sov. Phys. Semiconductors USSR* **16** 772

[10] Yoffe A D 2001 *Adv. Phys.* **50** 1

[11] Lin H B, Eversole J D and Campillo A J 1992 *J. Opt. Soc. Amer. B: Opt. Phys.* **9** 43

[12] Balachandran R M, Pacheo D P and Lawandy N M Appl. Optics. **35** 640 1996

[13] Lawandy N M 1995 'Optical gain medium having doped nanocrystals of semiconductors of semiconductors and also optical scatterers' US Patent 5,434,878

[14] Wooten F 1972 *Optical Properties of Solids* (Academic Press)

[15] Bohren C E and Huffmann D R 1983 *Absorption and Scattering of Light by Small Particles* (Wiley)

[16] Link S and El-Sayed M A 1999 *J. Phys. Chem. B* **103** 8419

[17] Sheng P 1980 *Phys. Rev. Lett.* **45** 60

[18] Giaever I 1973 *J. Immunol.* **110** 1424

[19] Biacore™ *http://www.biacore.com*

[20] Kohler J M, Csaki A, Reichert J, Moller R, Straub W and Fritzsche W 2001 *Sensors and Actuators B: Chemical* **76** 166

[21] He L, Musick M D, Nicewarner S R, Salinas F G, Benkovic S J, Natan M J and Keating C D 2000 *J. Am. Chem Soc.* **122** 9071

[22] Edelstein A S and Cammarata R C 1996 *Nanoparticles: Synthesis, Properties and Applications* (Bristol: Institute of Physics Publishing)

[23] Woggon U 1997 *Optical Properties of Semiconductor Quantum Dots* (Berlin: Springer)

[24] Colvin V L, Schlamp M C and Alivisatos A P 1994 *Nature* **370** 354

[25] Salata O V, Dobson P J, Sabesan S, Hull P J and Hutchison J L 1996 *Thin Solid Films* **288** 235

[26] Cotter D, Girdlestone H P and Moulding K 1991 *Appl. Phys. Lett.* **58** 1455

[27] Empedocles S A and Bawendi M G 1997 *Science* **278** 2114

[28] Miller D A B, Chemla D S and Schmitt-Rink S 1986 *Phys. Rev. B* **33** 6976

[29] Whitehead M, Parry G, Woodbridge K, Dobson P J and Duggan G 1988 *Appl. Phys. Lett.* **52** 345

[30] Erley G, Gorer S and Penner R M 1998 *Appl. Phys. Lett.* **72** 2301

[31] Murray C B, Norris D J and Bawendi M G 1993 *J. Amer. Chem. Soc.* **115** 8706

[32] Danek M, Jensen K F, Murray C B and Bawendi M G 1994 *Appl. Phys. Lett.* **65** 2795

[33] Shim M, Wang C and Guyot-Sionnest P 2001 *J. Phys. Chem.* **105** 2369

[34] Kanatzidis M G 1990 *Chemical and Engineering News* p 36 December

[35] J W Ostrander, A A Mamedov and N A Kotov 2001 *J. Am. Chem Soc.* **123** 1101

[36] Taylor R, Dobson P J and Hutchison J L unpublished work

[37] Wakefield G and Dobson P J unpublished work

[38] Phosphor nanoparticles can be purchased: *http://www.oxonica.com*

[39] Gledhill S E, Kaufmann C, Neve S, Dobson P J and Hutchison J L 2001 'A comparison of Zn(OH,S) nanoparticle thin films prepared by colloidal and non-colloidal routes' MRS Spring Meeting, San Francisco, April

[40] Schmidt H 2000 *Mater. Sci. Technol.* **16** 1356

[41] Aspnes D E 1982 *Am. J. Phys.* **50** 704

[42] Aspnes D E 1982 *Phys. Rev. Lett.* **48** 1629

[43] Niu F 1998 'Functional nanocomposite thin films by co-sputtering' DPhil thesis, University of Oxford

Chapter 11

Nanocrystallization in steels by heavy deformation

Minoro Umemoto

Introduction

This chapter reviews nanocrystallization in steels by intensive plastic deformation, focused mainly on ball milling. Nanocrystallization by ball milling occurs in the following steps. In the early stages of ball milling, work-hardening occurs, the dislocation density increases and a cellular structure develops. In the middle stages of ball milling, a drastic transition from work-hardened to layered nanostructures takes place near the powder surfaces. The layered nanostructure has a well developed granular structure with thickness less than 100 nm and is almost dislocation free. A clear boundary exists between the layered nanostructure and the work-hardened regions. It has been suggested that when the dislocation cell size reaches a small critical value as a consequence of the severe and high strain rate deformation, continuous recrystallization takes place and results in the layered nanocrystalline structure. By further milling, the thickness of the layered nanostructure decreases and the layers become sub-divided by the rotation of equiaxed region. In the final stages of ball milling, randomly oriented equiaxed regions are produced, containing about 10 nm diameter grains. By annealing, recrystallization does not take place in the nanocrystalline ferrite region and instead slow grain growth takes place.

Background

Nanocrystalline materials have attracted considerable scientific interest in the past decade [1, 2]. Because of the high density of grain boundaries and large number of ultrafine grains, nanocrystalline materials posses fundamentally different properties from their conventional coarse-grained polycrystalline

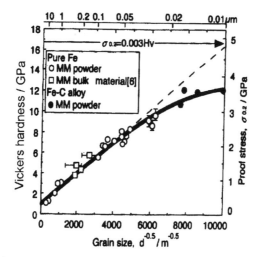

Figure 11.1. Hardness as a function of grain size in steels [3].

counterparts, which may find potentially important technological applications.

It has been well known that yield stress increases with the decrease in grain size according to the Hall–Petch relationship

$$\sigma = \sigma_0 + k_y d^{-0.5} \tag{11.1}$$

where σ_0 and k_y are constants and d is the grain diameter. Figure 11.1 [3] shows this relationship in steels. The increase in strength with $d^{-0.5}$ is observed in ferrite grains down to 50 nm. The deviation from the Hall–Petch relationship for grain diameters less than 50 nm is probably due to grain boundary sliding. It is also well known that grain refinement lowers the ductile–brittle transition temperature in steels linearly with $d^{-0.5}$.

Steel companies have made great efforts to refine grain size to increase both the strength and toughness of steels. Thermomechanical control processing (TMCP), which consists of controlled rolling and accelerated cooling, has been developed to produce steels with ferrite (α) grain diameters down to 5 μm. In order to obtain finer grains with sizes less than 1 μm in steels extended thermomechanical processes have been developed. The typical thermomechanical processes developed recently in Japan are shown in figure 11.2 [4]. The refinement of α grain size to around 1 μm has been achieved by heavy deformation either (a) in the supercooled austenite (γ) region through inducing low temperature diffusive transformation or (b) in the α region through inducing low temperature recrystallization. Spontaneous reverse transformation has also been demonstrated to result in very fine γ grain size of less than 1 μm by heavy deformation just below A_1 temperature, as shown in figure 11.2(c) [5].

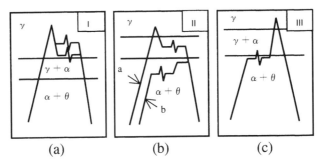

Figure 11.2. Heating patterns in thermomechanical processes developed in the Super Metal Project [4].

It is known that the microstructure after heavy deformation, for example by cold rolling or drawing [6], consists of subgrains characterized by dislocation cell boundaries with low angle misorientations. Ultrafine-grained structures with granular and high angle boundaries are obtained when very large deformation at low temperatures is applied. Some of these processes are shown in figure 11.3. These are (a) severe torsion straining under high pressure [7–12], (b) equal channel angular pressing [13–18], (c) ball milling [19–22] and (d) accumulative roll-bonding [23, 24].

The present chapter discusses microstructure evolution in steels by ball milling, developing a nanocrystalline state. With the progress of ball milling the work-hardened ferrite discontinuously changes to a layered nanostructure.

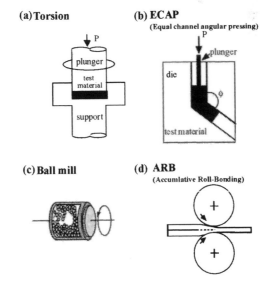

Figure 11.3. Heavy deformation processes developed to induce nanocrystallization.

A critical dislocation density is needed to induce this transition, and a high strain rate may be a necessary condition. This is demonstrated by ball drop testing, in which a ball with a weight is dropped on to a flat surface bulk specimen.

Nanocrystallization in steels by ball milling

Ball milling has been used to produce nanostructured materials since 1988. Shingu *et al* [19] first observed the formation of nanocrystallized Al-Fe composite by horizontal ball milling, and Jang and Koch [20] reported the nanocrystallization of pure iron by vibrational milling. Since then, nanocrystallization by ball milling in various metal elements with bcc and hcp [21] and fcc [22] structures have been studied.

A typical scanning electron micrograph of ball milled Fe-0.1C alloy is shown in figure 11.4(a). Two types of structure are seen: a uniform layered structure near the surface of the powder (dark smooth contrast), and a deformed structure in the interior of the powder (bright contrast). The boundaries between these two types of structure are clear and sharp as

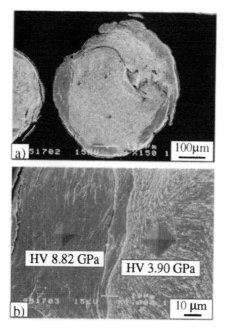

Figure 11.4. Scanning electron micrographs of Fe-0.10C martensite after ball milling for 360 ks. (a) Cross section of a powder, (b) enlargement of the boundary between the work-hardened and layered nanostructure regions.

shown in figure 11.4(b). The uniform layered structure near the surface of the powder is nanostructured, as revealed by transmission electron microscopy. These two types of structure are observed in all kinds of steels after ball milling, irrespective of the carbon content (0–0.9 wt% C) or starting microstructure (ferrite, martensite, pearlite or spheroidite) [25–29].

A drastic change in hardness is seen across the two regions, as shown in figure 11.5. The microhardness of the nanocrystalline ferrite region in pure iron is about 7.5 GPa, almost double the microhardness of 3.5 GPa in the work-hardened region. The evolution of microhardness with ball milling time in pure iron is shown in figure 11.6 [28]. The microhardness of the work-hardened region increases with milling time gradually from 1.3 GPa initially to about 3.1 GPa after 720 ks milling. In contrast, the hardness of the layered nanostructure region starts from 5.4 GPa after 180 ks milling, and reaches a saturated value of 8.2 GPa after 1800 ks milling. Hardnesses between 3.1 and 5.4 GPa are not observed. The hardening of the work-hardened region is attributed to dislocation hardening, and the hardening of the layered nanostructure is attributed to grain refinement.

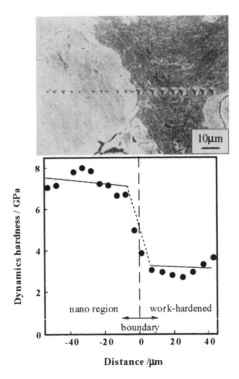

Figure 11.5. Microstructure and dynamic microhardness profile across the boundary of work-hardened and layered nanostructure regions in the 360 ks ball milled Fe-0.03C sample.

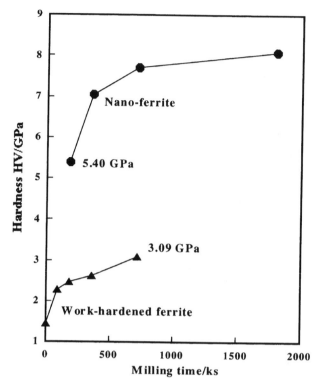

Figure 11.6. Microhardness change as a function of milling time in work-hardened and nanocrystalline ferrite regions in an Fe-0.004C sample.

Detailed transmission electron microscope observations of microstructural evolution during ball milling are shown in figures 11.7–11.11. Figure 11.7(a) shows a work-hardened region in pure iron milled for 360 ks [28]. Dislocation cells are formed, with an average size of about 200 nm. Figure 11.7(b) shows a layered nanostructure region near the surface of the same powder [28]. Elongated grains are formed, with an average thickness of less than 100 nm. Figure 11.8 shows a high magnification image of the layered nanostructure. High angle grain boundaries are well developed and there are almost no dislocations in the interior of the grains. Figure 11.9 shows the interface between the work-hardened and layered nanostructure regions in an Fe-0.004C alloy after milling for 360 ks. The layered nanocrystalline ferrite has an average thickness of about 100 nm in the lower part of the micrograph, and a dislocation cell structure in the upper part of the micrograph. The change in structure takes place without any detectable transition region. Figure 11.10 shows a high magnification image of the layered nanostructure close to the powder surface of an Fe-0.89C alloy with a spheroidite structure after milling for 360 ks. The nanocrystalline

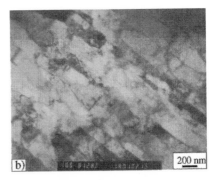

Figure 11.7. Transmission electron micrographs of Fe-0.004C milled for 360 ks. (a) work-hardened region near surface and (b) nanocrystalline region at interior area of a powder.

layers subdivide into fine equiaxed regions with different orientations. This indicates that the rotation of small sub-grains takes place in the layered nanostructure, and equiaxed nanocrystalline ferrite grains start to form. After milling for 1800 ks, uniformly distributed fine grains are formed, about 5–10 nm in size (figure 11.11). Grains are randomly oriented, as seen from both the continuous diffraction rings and the lattice fringes observed in by high-resolution electron microscopy. The grain boundaries are partly disordered, with complex contrast.

Transmission electron microscope observations of microstructural evolution by ball milling are summarized schematically in figure 11.12. In the

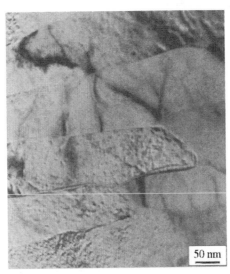

Figure 11.8. Transmission electron micrograph showing the nanostructure region in Fe-0.004C after 360 ks ball milling.

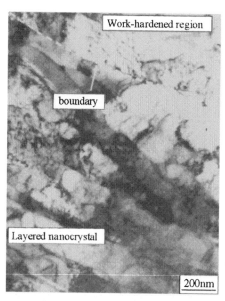

Figure 11.9. Transmission electron micrograph showing the boundary between the work-hardened and layered nanostructure regions in Fe-0.004C spheroidite after 360 ks ball milling.

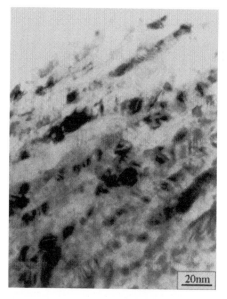

Figure 11.10. High-resolution transmission electron micrograph showing the misoriented small regions in the layered nanostructure in Fe-0.89C spheroidite after 360 ks ball milling.

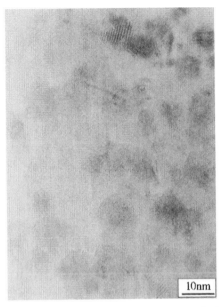

Figure 11.11. High-resolution transmission electron micrograph showing uniformly distributed 5–10 nm ferrite grains in Fe-0.89C spheroidite after 1800 ks ball milling.

initial stage of ball milling, dislocation density increases and a cell structure is developed with milling time. With further milling, the cellular structure transforms to a granular one, and layered ferrite with a thickness of several tens of nanometres is formed. Further milling leads to the rotation

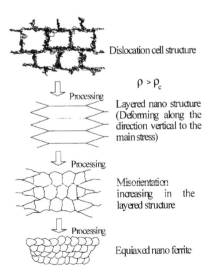

Figure 11.12. A schematic drawing of the microstructure evolution by ball milling.

of local regions in the microstructure. Finally randomly-oriented nanometre-sized equiaxed grains are formed.

Annealing behaviour

Annealing leads to structural changes in nanocrystalline ferrite regions which are quite different from work-hardened regions. Figure 11.13 shows the microstructure changes as a function of annealing temperature in pure iron ball milled for 360 ks and then heat treated for 3.6 ks at 673, 873 and 1073 K [28]. After annealing at 673 K, as shown in figure 11.13(a), the work-hardened region in the lower right area develops equiaxed grains with an average size of about 0.5 μm as the result of recovery, but the nanocrystalline ferrite region in the upper right area shows almost no detectable change. After annealing at 873 K, as shown in figure 11.13(b), the

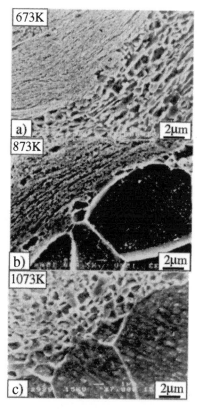

Figure 11.13. Scanning electron micrographs of Fe-0.004C milled for 360 ks and annealed for 3.6 ks at (a) 673 K, (b) 873 K and (c) 1073 K.

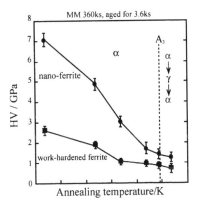

Figure 11.14. Microhardness evolution of the two types of regions in the 360 ks ball milled Fe-0.004C as a function of annealing temperature (for 3.6 ks).

recrystallized grains in the work-hardened region increase in size to 10 μm, while the nanocrystalline ferrite region remains almost unchanged. Further increase in the annealing temperature to 1073 K, as shown in figure 11.13(c), leads to the onset of detectable grain growth in the nanocrystalline ferrite region, with the grain size increasing to about 1 μm. The change in microhardness of the two types of region is shown in figure 11.14 as a function of annealing temperature [28]. Nanocrystalline ferrite regions have a higher hardness than work-hardened regions, although the difference becomes smaller with increasing annealing temperature. Nanocrystalline ferrite regions show a gradual decrease in hardness while work-hardened regions show sharp softening at around 673 K, caused by recrystallization.

The difference in annealing behaviour in the two regions can be explained as follows. In work-hardened regions, conventional recrystallization takes place driven by a high dislocation density. In nanocrystalline ferrite regions, recrystallization does not occur since there are almost no dislocations in the grain interiors, and gradual grain growth takes place instead. It seems that grain growth in nanocrystalline ferrite is substantially slower than that in a coarse grained counterpart. The slow grain growth rate has been observed in various nanocrystalline materials and it has been suggested that the low mobility of triple junction is responsible [30, 31].

Nanocrystallization in ball drop tests

Microstructural evolution during ball milling has been intensively studied, as mentioned above. However, in ball milling, the deformation mode is very complicated and it is hard to avoid contamination of the powders from the containing vial or the atmosphere. Thus the deformation conditions to

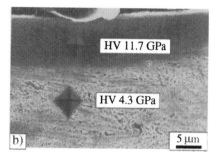

Figure 11.15. Scanning electron micrographs of Fe-0.89C near the surface after ball drop test. (a) spheroidite and (b) pearlite structure.

induce nanocrystallization, such as the strain rate, the amount of strain in each collision and the number of collisions, are unknown. Nanocrystalline structures similar to those produced by ball milling can also be obtained by ball drop testing. A heavy ball dropped on to a bulk specimen can produce surface deformation with a true strain of around 2 and a high strain rate of 10^4/s, similar to those achieved in ball milling.

Figure 11.15 [32] shows typical microstructures observed in a eutectoid steel with (a) spheroidite and (b) pearlite structures respectively. A uniform dark layer with a thickness of about 15 μm is seen at the steel surface. The surface hardness is very high, cementite particles in the microstructure are mostly dissolved, and the formation of about 100 nm sized nanocrystalline ferrite is confirmed by transmission electron microscopy. Typical test conditions to obtain such a nanocrystalline surface layer are 6 mm ball diameter, 3 kg weight, and three ball drops from 1 m height. Even a single drop leads to the formation of nanocrystalline ferrite when the specimen is predeformed by rolling.

A high strain rate deformation of around 10^4/s is important to produce a ferrite nanostructure. The ball drop test confirms that for nanocrystallization by ball milling deformation plays the major role but contamination has a minor effect. The dislocation density at a given degree of deformation increases with increasing strain rate. When the strain rate increases above a critical value, the dislocation density reaches a corresponding critical value and induces a transition from a dislocation cellular structure to a granular structure. This is known as continuous recrystallization. When deformation is applied at room temperature in steels, the grains produced by continuous recrystallization are smaller than 100 nm in size.

Nanocrystallization mechanism

Various heavy or severe plastic deformation methods have been proposed to produce nanostructured materials, such as torsion straining under high pressure [7–12], equal channel angular pressing (ECAP) [13–18], multiple forging [33], accumulated roll-bonding [23, 24], ball milling [19–22], shot peening [34] and so on. Among these, ball milling has been reported to produce the smallest grain sizes of around 10 nm.

Nanocrystalline steels are quite brittle, and it is difficult to obtain nano-structured bulk materials by severe plastic deformation. Severe plastic deformation of aluminium alloys is discussed in chapter 20. So-called bulk nanostructured steels reported so far [12] seem to contain both work-hardened regions with some ductility and brittle nanocrystalline regions. A random mixture of two types of region in bulk nanostructured steels makes it difficult to determine the mechanism of nanostructure formation. In studies of ball milling or ball drop testing, the degree of deformation and the strain rate are both functions of the depth below the surface. This makes it possible to observe the effect of deformation and strain rate on the formation of nanostructures as a function of distance from the surface. From ball milling and ball drop testing, it is clear that the transition from a cellular dislocation structure to a nanosized granular structure is a discontinuous process, with a critical deformed density for the transition to take place [25–29].

The most important point of the nanocrystallization mechanism by severe deformation is the transition from a cellular dislocation structure to a nanocrystalline granular structure [35]. Valiev *et al* [11] proposed a model of nanocrystallization, as shown in figure 11.16. When the dislocation density in the cell walls achieves some critical value, as shown in figure 11.16(a), an annihilation of dislocations of different signs occurs at the cell boundaries, as shown in figure 11.16(b). As a result, excess dislocations of single sign remain, as shown in figure 11.16(c). Excess dislocations with Burgers vectors perpendicular to the cell boundaries lead to an increase in misorientation and when their density rises they cause the transition to a granular structure. This model seems reasonable to explain the structural changes observed during ball milling or ball drop experiments. However, the detailed atomistic model of the transition from dislocation cells to grain boundaries remains still to be solved.

In conventional recrystallization processes (called discontinuous recrystallization), the recrystallization nuclei are considered to be small grains

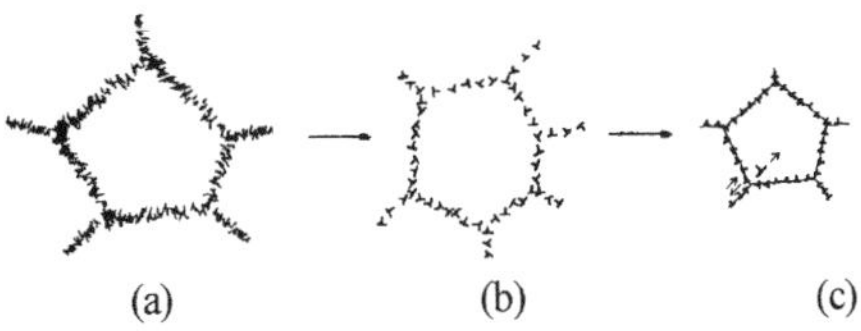

(a)   (b)   (c)

Figure 11.16. Schematic model for the evolution of dislocation structures at different stages during intense plastic straining [11].

without dislocation formed from the original grain boundaries or the smallest dislocation cells with high misorientations (larger than 15°) with neighbouring grains. In the latter case, the cellular dislocation structure changes to a granular structure by thermally assisted or thermally and stress assisted processes. When severe plastic deformation is applied, the cell size reaches the smallest critical size and the dislocation density in the cell walls reaches the critical density to make a transition to a granular structure. The dislocation cells in the whole specimen transform to a granular structure simultaneously. It seems that high strain rate raises the specimen temperature and a thermally assisted process may occur to some extent. Thus nanocrystallization by deformation can be considered to be a kind of continuous recrystallization [36] in which the transition from cells to granular structure occurs throughout the material by deformation.

The reason for the slow grain growth rate in nanostructured material remains unclear [30, 31]. If grain growth took place in a similar way in nanocrystalline and conventional coarse-grained materials, then very rapid grain growth during annealing would destroy the superior properties of nanostructured materials. It has been reported that a number of nanostructured materials exhibit abnormally small growth rate and this is also confirmed in steels. This makes nanostructured materials more useful. The mechanism of slow grain growth is not clear. Triple junctions show poor mobility and stable grain structures seem to be formed during grain growth in nanosized grains. In conventional grain growth, the grain size ratio before and after grain growth is usually below 5, much smaller than in nanostructured materials. This may provide insufficient reconstruction to allow the formation of such stable grain structures. Detailed studies of grain boundary reconstruction and rearrangement is needed.

Summary

This chapter has reviewed nanocrystallization in steels by severe plastic deformation, concentrating particularly on ball milling. The evolution of microstructure during nanocrystallization by severe plastic deformation can be divided into three stages. The first stage is an increase in the dislocation density and the formation of a cellular structure. The second stage is the transition from a cellular to a layered structure, which probably occurs when the dislocation density at the cell walls reaches a critical value. The layers are nanostructured, are less than 100 nm thick, and are almost dislocation free in the interiors of the grains. In ball milling or during ball drop testing, the transition from a cellular to a layered structure occurs drastically without a detectable transition region. The final stage is the formation of randomly oriented equiaxed 10 nm sized grains with large misorientations in orientation. The equiaxed grains form from the layered nanostructure by increasing local misorientations.

The unique nature of nanostructured steel is also seen in its annealing behaviour. Recrystallization does not take place in the nanocrystalline ferrite because there are no dislocations in the interiors of the grains. Grain growth in the nanocrystalline ferrite is quite slow, for reasons which have not yet been established fully.

Ball drop tests can produce structural evolution similar to ball milling. Even with a single test, nanocrystalline ferrite forms. High strain rate deformation at about 10^4/s is clearly an important method of producing nanocrystalline ferrite. A large number of repeated deformations in ball milling can produce nanostructured material, as also can a small number of deformations with large strain and high strain rate in ball testing.

References

[1] Suryanarayana C 1995 *Int. Mater. Rev.* **40** 41

[2] Gleiter H 2000 *Acta Mater. Sci.* **48** 1

[3] Hidaka H, Suzuki T, Kimura Y and Takaki S 1999 *Mater. Sci. Forum* **304–306** 115

[4] Hagiwara Y 2001 *Proc. 3rd Symposium of Super Metal* p 13

[5] Yokota T, Shiraga T, Niikura M and Sato K 2000 *Proc. Int. Conf. on Ultrafine Grained Materials, Nashville* (TMS) p 267

[6] Langford G and Cohen M 1969 *Trans. ASM* **62** 623

[7] Valiev R Z, Kaibyshev O A, Kurnetsov R I, Musalimov R Sh and Tsenev N K 1988 *DAN SSSR* **301**(4) 864

[8] Valiev R Z, Korznikov A V and Mulyukov R R 1993 *Mater. Sci. Eng. A* **168** 141

[9] Korznikov A V, Ivanisenko Yu V, Laptionok D V, Safarov I M, Pilyugin V P and Valiv R Z 1994 *NanoStruct. Mater.* **4** 159

[10] Mishin O V, Gertsman V Yu, Valiev R Z and Gottstein G 1996 *Scripta Mater.* **35** 873

[11] Valiev R Z, Ivanisenko Yu V, Rauch E F and Baudelet B 1996 *Acta Mater.* **44** 4705

[12] Valiev R Z, Islamgaliev R K and Alexandrov I V 2000 *Prog. Mater. Sci.* **45** 103

[13] Segal V M, Reznikov V I, Drobyshevskiy A D and Kopylov V I 1981 *Russ. Metall.* **1** 99

[14] Valiev R Z, Krasilnikov N A and Tsenev N K 1991 *Mater. Sci. Eng. A* **137** 35

[15] Ahmadeev N H, Valiev R Z, Kopylov V I and Mulyukov R R 1992 *Russ. Metall.* **5** 96

[16] Valiev R Z, Korznikov A V and Mulyukov R R 1992 *Phys. Met. Metall.* **73** 373

[17] Shin D H, Kim B C, Kim Y-S and Park K-T 2000 *Acta Mater.* **48** 2247

[18] Shin D H, Kim B C, Park K-T and Choo W Y 2000 *Acta Mater.* **48** 3245

[19] Shingu P H, Huang B, Nishitani S R and Nasu S Suppl. 1988 *Trans. JIM* **29** 3

[20] Jang J S C and Koch C C 1990 *Scripta Metall.* **24** 1599

[21] Fecht H J, Hellstern E, Fu Z and Johnson W L 1990 *Met. Trans. A* **21** 2333

[22] Eckert J, Holzer J C, Krill C E III and Johnson W L 1992 *J. Mater. Res.* **7** 1751

[23] Saito Y, Tsuji N, Utsunomiya H, Sakai T and Hong R G 1998 *Scripta Mater.* **39** 1221

[24] Saito Y, Utsunomiya H, Tsuji N and Sakai T 1999 *Acta Mater.* **47** 579

[25] Umemoto M, Liu Z G, Hao X J, Masuyama K and Tsuchiya K 2001 *Mater. Sci. Forum* **360–362** 167

[26] Liu Z G, Hao X J, Masuyama K, Tsuchiya K, Umemoto M and Hao S M 2001 *Scripta Mater.* **44** 1775

[27] Umemoto M, Liu Z G, Masuyama K, Hao X J and Tsuchiya K 2001 *Scripta Mater.* **44** 1741
[28] Yin J, Umemoto M, Liu Z G and Tsuchiya K unpublished work
[29] Umemoto M, Liu Z G, Xu Y and Tsuchiya K unpublished work
[30] Moelle C H and Fecht H J 1995 *NanoStruct. Mater.* 6 421
[31] Marlow T R and Koch C C 1997 *Acta Mater.* **45** 2177
[32] Umemoto M, Haung B, Tsuchiya K and Suzuki N unpublished work
[33] Imayev R M, Imayev V M and Salishchev G A 1993 *J. Mater. Sci.* **28** 289
[34] Tao N R, Sui M L, Ku J and Ku L 1999 *NanoStruct. Mater.* **11** 433
[35] Smirnova N A, Levit V I, Pilyugin V I, Kuznetsov R I, Davydova L S and Sazonova V A 1986 *Fiz. Met. Metalloved* **61** 1170
[36] Tsuji N, Ueji R and Saito Y 2000 *Proc. 21st Riso Int. Symp. on Materials Science* p 607

Chapter 12

Severe plastic deformation

Minoru Furukawa, Zenji Horita and Terence Langdon

Introduction

Processing through the application of severe plastic deformation provides an opportunity for achieving very significant grain refinement in a wide range of materials. This chapter examines the major procedures for imposing severe plastic deformation and describes the processing procedure and representative results achieved using the process of ECAP (ECAP).

Background

Considerable interest has developed recently in preparing materials with ultrafine grain sizes. This interest arises as a consequence of the conventional Hall–Petch relationship which is expressed in the form [1, 2]

$$\sigma_y = \sigma_0 + k_y d^{-1/2} \tag{12.1}$$

where σ_y is the yield stress of a polycrystalline material, σ_0 is termed the friction stress, k_y is a yielding constant and d is the grain size of the material. Thus, the tensile strength of a material at ambient temperatures varies inversely with the square-root of the grain size so that a reduction in grain size leads to a higher tensile strength and there is a corresponding increase in the toughness of the material. In addition, it has been noted that, if these ultrafine grain sizes are stable at high temperatures where diffusion becomes reasonably rapid, it may be possible to achieve a superplastic forming capability at rapid strain rates [3]. This latter objective is attractive because the commercial superplastic forming industry is currently restricted to producing parts at relatively slow production rates, typically of the order of $\sim 10^{-3}\,\mathrm{s}^{-1}$, so that each separate component requires a production time from ~ 20 to ~ 30 min. Thus, the fabrication of ultrafine grain sizes that are stable at high temperatures provides a potential for expanding superplastic forming operations from the current

187

production of high-cost low-volume components for the aerospace and other industries to the production of low-cost high-volume components associated with the automotive and consumer product industries.

Several different techniques have been utilized to produce ultrafine grain sizes in metals including inert gas condensation, high-energy ball milling and sliding wear. These various techniques are based on the growth of structures at the atomic level or on the consolidation of ultrafine particles and they have the potential for producing small samples of materials with grain sizes at least within the nanometre range (<100 nm) and typically in the range of $\sim$10–30 nm. These procedures have an excellent potential for fabricating materials suitable for use in the electronics and computer industries. Nevertheless, these various procedures suffer from two serious disadvantages which limit their effectiveness in the processing of samples for large-scale structural applications. First, the materials fabricated by these techniques always contain at least a small amount of residual porosity which cannot be removed by appropriately adjusting the processing parameters. Second, the dimensions of the samples produced in this way are very small and no procedures have been developed to date for scaling the processes to produce large bulk samples.

As a consequence of these difficulties, much interest has developed instead in the processing of materials using various procedures that involve imposing severe plastic deformation [4]. The various techniques adopted using severe plastic deformation have the objective of using conventionally processed metals and applying a straining procedure so that the grains become refined below the submicrometre level and possibly even to the nanometre level. It has been shown that severe plastic deformation procedures are capable of producing, in a simple effective manner, materials having grain sizes that are very significantly smaller than those generally produced using conventional thermomechanical processing. Furthermore, thermomechanical processing requires the development of a unique thermal and mechanical processing path for each separate alloy whereas severe plastic deformation can be applied in essentially the same way to all materials. The development and application of severe plastic deformation techniques has now reached the level that the topic was the subject of a NATO Advanced Research Workshop held in Moscow, Russia, in August 1999 [5].

This chapter is designed to fulfil two objectives. First, it provides a brief overview of the major techniques of severe plastic deformation currently available for the production of ultrafine-grained materials. Secondly, it illustrates the application of one of these techniques and describes some representative results.

Severe plastic deformation

The principle of severe plastic deformation processing is that a sample is subjected to a very intense plastic strain without incurring any concomitant

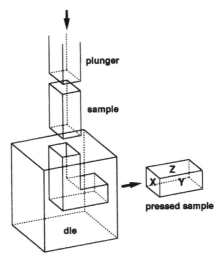

Figure 12.1. The principle of equal-channel angular pressing showing the three orthogonal planes *X*, *Y* and *Z*.

changes in the cross-sectional dimensions of the work-piece. Thus, severe plastic deformation differs in a very significant way from the more conventional and well-established metal-working processes such as rolling, extrusion, forging and drawing. An important additional requirement in successful severe plastic deformation is that the procedure introduces more than a simple redundant strain. For example, Richert and Richert [6] proposed in 1986 a metal-working process known as cyclic extrusion compression in which a sample is contained within a chamber and then extruded repeatedly backwards and forwards through a confining aperture. In principle, cyclic extrusion compression is capable of imposing an unlimited strain on any selected sample but the process is not satisfactory for the fabrication of ultrafine-grained materials because the strain introduced in the forward extrusion is effectively cancelled by the strain imposed in the reverse extrusion.

There are at the present time two major and distinct procedures for severe plastic deformation processing and these are illustrated schematically in figures 12.1 and 12.2. An additional processing method was also proposed very recently and this is illustrated in figure 12.3.

The first procedure, shown in figure 12.1, is known as equal-channel angular pressing (ECAP)* and it was developed initially by Segal *et al* [7] in the former Soviet Union as a technique for homogenizing cast billets. Subsequently, it was shown by Valiev *et al* [8] that this process may be used to produce materials with ultrafine grain sizes typically lying in the

---

* The same process is also termed equal-channel angular extrusion (ECAE) in the literature but, since there is no extrusion of the sample through a confining aperture, the acronym ECAP (equal-channel angular pressing) has been recommended for this processing method [5].

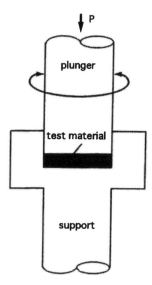

Figure 12.2. The principle of high-pressure torsion: *P* is the external applied pressure.

submicrometre range ($\sim$100 nm to 1 µm). In processing by equal-channel angular pressing, a die is constructed, usually from a tool steel, and this die contains a channel that is bent into an L-shaped configuration: for example, the channel is bent through an angle of 90° in figure 12.1. The test sample is machined so that it fits snugly within the channel and it is then pressed through the die using a plunger. In practice, a second sample is generally inserted into the die in order to successfully remove the first sample. It is apparent from the schematic illustration in figure 12.1 that the sample emerges from the die having experienced no change in the cross-sectional dimensions: the planes labelled *X*, *Y* and *Z* in figure 12.1 are the transverse, flow and longitudinal planes, respectively, and they correspond to the planes perpendicular to the direction of flow and parallel to the side and top faces at the point of exit from the die, respectively. In practice, there are end effects which have been examined experimentally [9] but, for convenience, are not depicted on the pressed sample in figure 12.1.

Since there is no change in the cross-section of the specimen in a single passage through the equal-channel angular pressing die, additional pressings may be conducted on the same sample in order to achieve very high total strains. Thus, there are numerous experimental parameters that may affect the microstructures developed through equal-channel angular pressing including the number of passes through the die (and thus the total imposed strain), whether or not the sample is rotated about the longitudinal axis between repetitive pressings, the temperature and the speed of pressing and the angle between the two parts of the channel within the equal-channel angular pressing die.

The second procedure, shown in figure 12.2, is high-pressure torsion (high-pressure torsion) which is based on a processing technique introduced initially by Smirnova *et al* [10]. In high-pressure torsion, the sample is in the form of a small disc, typically having a diameter of ~1 cm and a thickness of ~3 mm, and it is subjected to a high confining pressure and then strained in torsion. Experiments show that high-pressure torsion is especially effective for producing materials with extremely small grain sizes. Typically, the grain sizes introduced by high-pressure torsion are in the nanometre range (<100 nm) and therefore they are generally significantly smaller than those produced by equal-channel angular pressing. Nevertheless, this technique has the disadvantage that it utilizes specimens in the form of relatively small discs and there appears at the present time to be no simple method for scaling the process to incorporate larger samples. In the absence of any significant developments, it is reasonable to conclude that processing by high-pressure torsion is not viable for the production of large bulk materials. An additional problem with high-pressure torsion is that the microstructures produced during processing are dependent upon a number of factors including the magnitude of the applied pressure and the precise location within the disc [11].

A third processing method, illustrated in figure 12.3, was proposed recently by Zhu *et al* [12] and it is termed repetitive corrugation and straightening. The principle of this processing method is that, as in equal-channel angular pressing, a large amount of plastic deformation may be introduced into the sample without any change in the cross-sectional dimensions. Specifically, the work-piece is initially pressed into a corrugated shape by placing it between the two platens of a hydraulic press and using the loading configuration shown in figure 12.3 and secondly the work-piece is straightened by placing it in a corrugated form between the two flat platens. Again, very high strains may be imposed by repeating these two steps in a cyclic manner and, as in equal-channel angular pressing, the sample may be rotated between each corrugation-straightening cycle. Very little information is

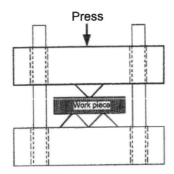

Figure 12.3. The principle of repetitive corrugation and straightening showing the procedure for introducing corrugation.

currently available on the microstructures produced using this new technique but there is experimental evidence that the grain size of high purity Cu may be effectively reduced through repetitive corrugation and straightening from ~760 µm to ~500 nm [13] thereby suggesting that the processing procedure has similarities to conventional equal-channel angular pressing.

It appears that the various severe plastic deformation processing procedures of equal-channel angular pressing, high-pressure torsion and repetitive corrugation and straightening all produce materials with ultrafine submicrometre grain sizes but high-pressure torsion processing is suitable only for small discs and the repetitive corrugation and straightening process is sufficiently new that there is very limited information regarding its efficiency. Accordingly, attention will be devoted exclusively in this report to the principles and microstructures associated with processing using equal-channel angular pressing.

Equal-channel angular pressing

When a sample passes through an equal-channel angular pressing die, it experiences simple shear at the shearing plane between the two parts of the channel on either side of the bend. In practice, the imposed strain is dependent both upon the angle Φ between the two parts of the channel and the angle Ψ which delineates the outer arc of curvature at the point where the two parts of the channel intersect. These two angles are defined explicitly in figure 12.4 which shows a section through the equal-channel angular pressing die and the passage of a single sample ahead of a plunger.

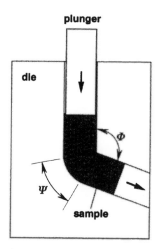

Figure 12.4. A section through an ECAP die showing the two internal angles Φ and Ψ.

It can be shown from first principles that the total strain imposed on a sample in a single pass through the die ε_1 is given by [14]

$$\varepsilon_1 = \frac{1}{\sqrt{3}} \left\{ 2 \cot \left(\frac{\Phi}{2} + \frac{\Psi}{2} \right) + \Phi \left(\frac{\Phi}{2} + \frac{\Psi}{2} \right) \right\}. \qquad (12.2)$$

For the situation where the same sample is pressed repetitively through the die for a total of N passes, the strain on a single pass ε_1 is then multiplied by N.

There is general consistency with the predictions of equation (12.2) in model experiments where layers of coloured plasticine were pressed through a Plexiglas die except only that the experiments revealed the possibility of frictional effects adjacent to the die walls [15]. Similar agreement has been reported also through pressing samples incorporating a grid pattern [16] and, except only near the sample edges because of frictional effects, through two-dimensional finite element modelling [17]. A slightly different relationship for ε_1 was suggested recently [18] but it can be shown that this alternative form is identical to equation (2) to within $<5\%$ under all possible pressing conditions for any angle of $\Phi \geq 90°$ [19].

The precise implications of equation (12.2) are illustrated in figure 12.5 where ε_1 for a single pass with $N = 1$ is plotted as a function of the angle between the two parts of the channel Φ for various values of the arc of curvature Ψ from $0°$ to $90°$ [20]. It is apparent from figure 12.5 that a typical equal-channel angular pressing die having $\Phi = 90°$ produces a strain close to ~ 1 for a single pass through the die and this magnitude of the strain is relatively insensitive to the value of Ψ.

When samples are pressed repetitively, they may be rotated about their longitudinal axes between consecutive passes so that different slip planes become activated [21]. In practice, it is possible to define four distinct

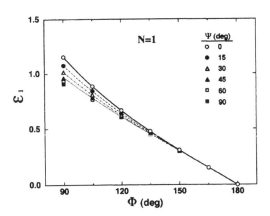

Figure 12.5. Variation of the imposed strain in equal-channel angular pressing with the internal channel angle Φ: the values for the arc of curvature, Ψ, range from $0°$ to $90°$.

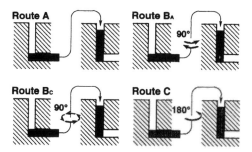

Figure 12.6. The four different processing routes in equal-channel angular pressing.

processing routes as illustrated schematically in figure 12.6, where route A denotes consecutive pressings without any rotation of the sample, routes B_A and B_C denote rotations by 90° between each pass either in alternate directions or in the same direction, respectively, and route C denotes rotations by 180° between consecutive passes [22].

In order to examine the shearing characteristics associated with these different processing routes, it is convenient to consider the deformation associated with the passage of a simple cubic element through an equal-channel angular pressing die having $\Phi = 90°$ and, for simplicity, $\Psi = 0°$ [23]. Figure 12.7 shows that the cubic element on the left is sheared into a rhombohedral shape as it passes through the theoretical shear plane shown shaded at the intersection of the two parts of the channel. Also included in figure 12.7 are illustrations of the macroscopic grain distortions (shown shaded) and the associated shearing planes within the grains (shown as straight lines) for each of the three planes of sectioning, designated as

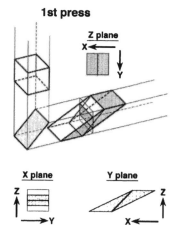

Figure 12.7. The deformation of a cubic element (at left) on the first pass through an equal-channel angular pressing die with $\Phi = 90°$ and $\Psi = 0°$.

planes X, Y and Z in figure 12.1. Thus, the grains are distorted on the X and Y planes in the first pass through the die but there is no corresponding distortion of the grains on the Z plane. The implications of figure 12.7 have been confirmed experimentally by using samples of pure aluminium and using optical microscopy to examine the surfaces after equal-channel angular pressing [24].

The situation for the second pass in equal-channel angular pressing is dependent upon whether the sample is rotated about the longitudinal axis between the first and second pass. Figure 12.8 illustrates the predictions for three separate situations: (a) for route A where there is no rotation,

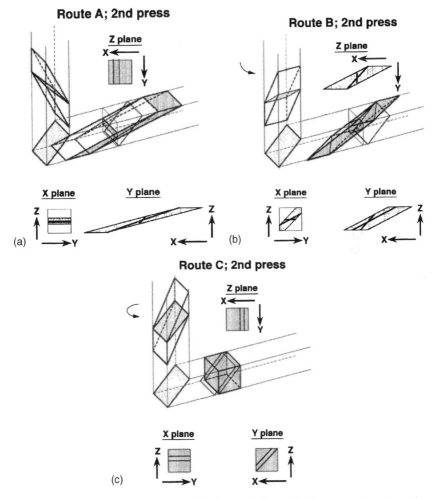

Figure 12.8. The second pass for the cubic element in figure 12.7 for processing when using (a) route A, (b) route B and (c) route C.

(b) for route B where there is a rotation of 90° and it is noted that routes B_A and B_C are equivalent for the second pass and (c) for route C where there is a rotation of 180°, where again the inserts illustrate the deformations associated with the X, Y and Z planes. Thus, route A in figure 12.8(a) gives an elongation of the grains in the Y plane at an angle of ~15° to the X axis, a further compression of the grains on the X plane but no distortions of the grains on the Z plane. In route B shown in figure 12.8(b) there is now an elongation of the grains on each of the orthogonal planes of sectioning. In route C shown in figure 12.8(c) the rotation of 180° serves to restore the cubic element after the second pass. These predictions are also consistent with experimental observations on samples of pure aluminium [24].

For additional numbers of passes, it is convenient to consider only the predictions when processing using route B_C since it has been shown that this route leads most expeditiously to an array of essentially equiaxed grains separated by grain boundaries having high angles of misorientation [25]. In addition, experiments have shown that route B_C is the optimum processing route for achieving the maximum superplastic ductilities at elevated temperatures [26]. It should be noted that the predictions for the other processing routes are documented in detail elsewhere [23]. Figure 12.9 illustrates the distortions of the cubic element during the third, fourth

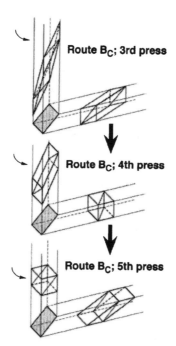

Figure 12.9. The 3rd, 4th and 5th passes for the cubic element in figure 12.7 when using processing route B_C.

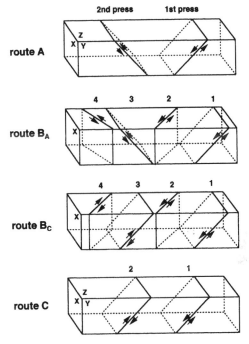

Figure 12.10. The shearing planes associated with the four different processing routes.

and fifth passes when using route B_C, thereby demonstrating that the cubic element is restored using route B_C after every $4n$ passes where n is an integer.

The precise shearing planes activated in each sample during pressing are illustrated schematically in figure 12.10 for the four different processing routes and for the X, Y and Z planes, where the shearing planes labelled 1–4 denote the occurrence of shearing on consecutive pressings through the equal-channel angular pressing die. Thus, route C entails repetitive shearings backwards and forwards on the same shearing plane, routes A and B_A give shearing on different sets of planes, and route B_C also entails shearing on four different planes but the cubic element is restored after four passes because it can be seen that there is effectively a cancellation of strains between both the first and third passes and the second and fourth passes.

The distortions of the grains associated with each of these four processing routes are illustrated schematically in figure 12.11 for the three orthogonal planes of sectioning and for a total of up to eight passes through the die. There is no deformation on the Z plane in route A, there is increasing deformation on each plane in route B_A, the cubic element is restored every $4n$ passes in route B_C with deformation on all three planes of sectioning, and the cubic element is restored every $2n$ passes in route C but without any corresponding deformation on the Z plane. This tabulation suggests that processing by route

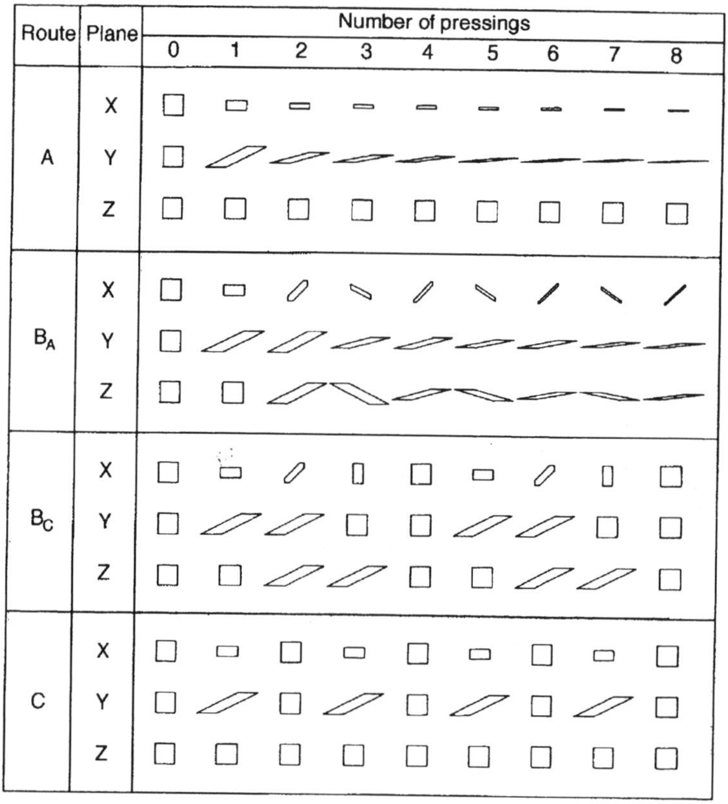

Route	Plane	Number of pressings								
		0	1	2	3	4	5	6	7	8
A	X									
	Y									
	Z									
B_A	X									
	Y									
	Z									
B_C	X									
	Y									
	Z									
C	X									
	Y									
	Z									

Figure 12.11. Tabulation showing the distortions of a cubic element for each processing route on each plane of sectioning up to a total of eight passes through the die.

B_C is the optimum procedure for achieving a uniform microstructure of equiaxed grains because the cubic element is periodically restored every four passes and deformation occurs on all three orthogonal planes. An additional reason favouring route B_C is the recent demonstration that shearing extends over larger angular ranges on the three orthogonal planes when using this processing route [27].

Microstructures and properties

Processing by equal-channel angular pressing generally reduces the grain size of polycrystalline materials to the micrometre or submicrometre level with the precise grain size dependent primarily upon the constituents within the material. For example, the grain size achieved after equal-channel angular pressing of pure aluminium at room temperature is approximately 1.3 μm

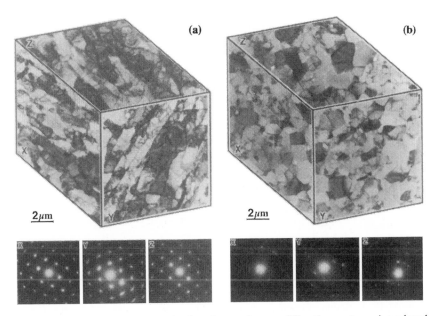

Figure 12.12. Microstructures and selected area electron diffraction patterns introduced into pure Al after equal-channel angular pressing at room temperature using processing route B_C for (a) one pass and (b) four passes.

[20, 28] but this average grain size is reduced to ~0.45μm with the addition of 1% Mg in solid solution and there is a further reduction to ~0.27 μm with the addition of 3% Mg [29]. The reduction in grain size associated with the addition of Mg in solid solution is attributed to the decreasing rate of recovery in the alloys.

Examples of the microstructures produced by equal-channel angular pressing are shown in figure 12.12 for samples of pure aluminium processed through a total of (a) one pass and (b) four passes, respectively [28]. Separate microstructures are shown in figure 12.12 for the X, Y and Z planes, the selected area electron diffraction (SAED) patterns were recorded for each plane using an aperture diameter of 12.3 μm and the samples were pressed at room temperature using an equal-channel angular pressing die with an internal angle of $\Phi = 90°$ so that the imposed strains are ~1 and ~4 in figures 12.12(a) and (b), respectively. It appears from figure 12.12(a) that the first pass through the die introduces an array of subgrains where, based on the selected area electron diffraction patterns, the individual grains are separated by boundaries having low angles of misorientation. These subgrains are elongated and lie in bands oriented essentially parallel to the top and bottom edges of the sample when viewed in the X plane, close to the shearing direction at 45° to the top and bottom edges in the Y plane and perpendicular to the direction of pressing in the Z plane: these

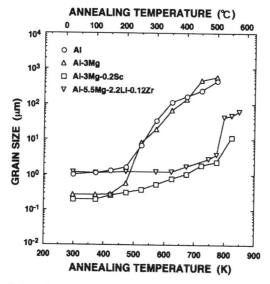

Figure 12.13. Variation of grain size with annealing temperature for four materials after equal-channel angular pressing.

results are therefore consistent with the predictions of the model presented in figure 12.7. After four passes in figure 12.12(b), the microstructure is essentially equiaxed on all three orthogonal planes of sectioning, there is no evidence for the occurrence of bands, and the selected area electron diffraction patterns suggest many of the boundaries now have high angles of misorientation. It is apparent that the boundaries are also well-defined and the grains contain a relatively small number of dislocations. Measurements show the average grain size in figure 12.12(b) is ~1.3 μm and this is in sharp contrast to the initial grain size in the unpressed material of ~1.0 mm. Thus, processing by equal-channel angular pressing at room temperature, even through a relatively small number of passes, leads to a remarkable refinement in the grain size of the material.

An additional important parameter in equal-channel angular pressing relates to the stability of these ultrafine grain sizes when the as-pressed samples are heated to elevated temperatures. Figure 12.13 shows the variation of the grain size with the annealing temperature for samples fabricated using equal-channel angular pressing and then annealed for 1 h at selected elevated temperatures [30–32]. It is apparent from this plot that the ultrafine grains introduced through equal-channel angular pressing are not stable at temperatures above *sim*500 K in samples of pure Al and the Al-3% Mg solid solution alloy whereas in the Al-Mg-Sc and Al-Mg-Li-Zr alloys, where precipitates are present, the grain sizes remain reasonably small and below 10 μm at temperatures up to 750 K. The retention of ultrafine grain

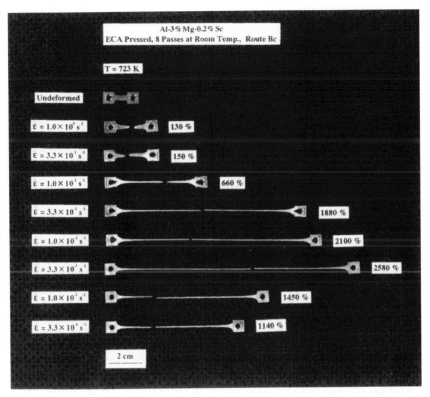

Figure 12.14. Examples of superplastic elongations in an Al-Mg-Sc alloy after processing by equal-channel angular pressing.

sizes in these two alloys is due to the presence of dispersions of fine Al_3Sc precipitates in the Al-Mg-Sc alloy and fine β'-Al_3Zr precipitates in the Al-Mg-Li-Zr alloy. It is reasonable to conclude, therefore, that these two alloys may have a potential for exhibiting superplastic ductilities at elevated temperatures.

Superplastic elongations are expected to occur in these aluminium-based alloys at temperatures above $\sim$550 K where diffusion is fairly rapid. Figure 12.14 shows an example of the high superplastic ductilities achieved in the Al-Mg-Sc alloy after tensile testing at a temperature of 723 K. The upper sample is untested and the other samples were processed by equal-channel angular pressing to eight passes and an imposed strain of $\sim$8 using route B_C at room temperature and they were then pulled to failure at initial strain rates ranging from $3.3 \times 10^{-4}\,\text{s}^{-1}$ (bottom) to $1.0\,\text{s}^{-1}$ (top) [33]. It is apparent that this material is exceptionally superplastic after equal-channel angular pressing with a maximum elongation at 723 K of $\sim$2580% when using an initial strain rate of $3.3 \times 10^{-3}\,\text{s}^{-1}$. Furthermore, the elongations achieved after equal-channel angular pressing are significantly higher than

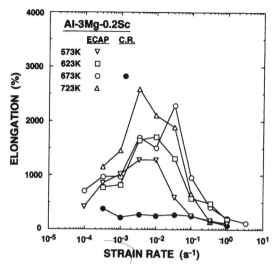

Figure 12.15. Variation of the elongation to failure with the initial strain rate for an Al-Mg-Sc alloy after ECAP and, for comparison purposes, after cold rolling (C.R.).

those achieved with the same alloy after conventional cold rolling. This difference is illustrated in figure 12.15 where the measured elongations to failure are plotted against the initial testing strain rate for samples subjected to equal-channel angular pressing and then pulled to failure at temperatures from 573 to 723 K and for samples subjected only to cold rolling at room temperature to an equivalent strain of ~2.4 and then tested in tension at 673 K [33]. For the cold-rolled alloy, the elongations are low and of the order of ~200% over the entire range of strain rates. This low ductility is due to the relatively low total strain of ~2.4 produced by cold rolling and the consequent presence of arrays of subgrain boundaries having low angles of misorientation which are not conducive to the occurrence of easy grain boundary sliding which is a necessary prerequisite for superplastic deformation.

Concluding remarks

Processing by severe plastic deformation is relatively new and it is an attractive and simple procedure for attaining very substantial grain refinement in a wide range of materials. It has significant advantages over conventional thermomechanical processing because essentially the same severe plastic deformation technique may be applied to all materials. The grain sizes introduced by severe plastic deformation are usually within the range from the nanometre to the low micrometre level. Sufficient data are now available to show that high-pressure torsion generally produces samples having grain sizes of <100 nm whereas equal-channel angular pressing typically produces materials with grain sizes in the range of ∼100 nm to ∼1 μm. Nevertheless, equal-channel angular pressing seems at present to be the optimum procedure for producing reasonably large bulk samples suitable for use in industrial applications.

Several recent developments have suggested new and important avenues for future research. These include combining equal-channel angular pressing for grain refinement with subsequent cold extrusion in order to optimize the strength of the as-pressed samples [34] and using cold rolling after equal-channel angular pressing to produce the ultrafine-grained material in a sheet form for subsequent superplastic forming operations [35]. There have been new developments in the application of the equal-channel angular pressing technique concentrating primarily on the necessity to achieve high total strains without removing the sample from the die. These procedures include the development of a multi-pass facility where a high strain is introduced in a single passage through the die [36] and the use of a rotary die where a high strain is achieved by rotating the die through 90° between consecutive pressings with the sample remaining *in situ* [37, 38]. The major interest now evident in the field of severe plastic deformation suggests these various techniques will be further developed to provide important alternative processing procedures for rapidly achieving materials with ultrafine-grained microstructures.

References

[1] Hall E O 1951 *Proc. Roy. Soc. B* **64** 747
[2] Petch N J 1953 *J. Iron Steel Inst.* **174** 25
[3] Ma Y, Furukawa M, Horita Z, Nemoto M, Valiev R Z and Langdon T G 1996 *Mater. Trans. JIM* **37** 336
[4] Valiev R Z, Islamgaliev R K and Alexandrov I V 2000 *Prog. Mater. Sci.* **45** 103
[5] Lowe T C and Valiev R Z (eds) 2000 *Investigations and Applications of Severe Plastic Deformation* (Dordrecht, The Netherlands: Kluwer)
[6] Richert J and Richert M 1986 *Aluminium* **62** 604
[7] Segal V M, Reznikov V I, Drobyshevskiy A E and Kopylov V I 1981 *Russian Metall.* **1** 99
[8] Valiev R Z, Krasilnikov N A and Tsenev N K 1991 *Mater. Sci. Eng. A* **137** 35
[9] Bowen J R, Gholinia A, Roberts S M and Prangnell P B 2000 *Mater. Sci. Eng. A* **287** 87
[10] Smirnova N A, Levit V I, Pilyugin V I, Kuznetsov R I, Davydova L S and Sazonova V A 1986 *Fiz. Met. Metalloved.* **61** 1170
[11] Zhilyaev A P, Lee S, Nurislamova G V, Valiev R Z and Langdon T G 2001 *Scripta Mater.* **44** 2753
[12] Zhu Y T, Jiang H, Huang J and Lowe T C 2001 *Metall. Mater. Trans.* **32A** 1559

[13] Huang J Y, Zhu Y T, Jiang H and Lowe T C 2001 *Acta Mater.* **49** 1497
[14] Iwahashi Y, Wang J, Horita Z, Nemoto M and Langdon T G 1996 *Scripta Mater.* **35** 143
[15] Wu Y and Baker I 1997 *Scripta Mater.* **37** 437
[16] Shan A, Moon I-G, Ko H-S and Park J-W 1999 *Scripta Mater.* **41** 355
[17] DeLo D P and Semiatin S L 1999 *Metall. Mater. Trans.* **30A** 1391
[18] Goforth R E, Hartwig K T and Cornwall L R 2000 *Investigations and Applications of Severe Plastic Deformation* ed. T C Lowe and R Z Valiev (Dordrecht, The Netherlands: Kluwer) p 3
[19] T Aida, K Matsuki, Z Horita and T G Langdon 2001 *Scripta Mater.* **44** 575
[20] Iwahashi Y, Horita Z, Nemoto M and Langdon T G 1997 *Acta Mater.* **45** 4733
[21] Segal V M 1995 *Mater. Sci. Eng. A* **197** 157
[22] Furukawa M, Iwahashi Y, Horita Z, Nemoto M and Langdon T G 1998 *Mater. Sci. Eng. A* **257** 328
[23] Furukawa M, Horita Z, Nemoto M and Langdon T G 2001 *J. Mater. Sci.* **36** 2835
[24] Iwahashi Y, Furukawa M, Horita Z, Nemoto M and Langdon T G 1998 *Metall. Mater. Trans.* **29A** 2245
[25] Oh-ishi K, Horita Z, Furukawa M, Nemoto M and Langdon T G 1998 *Metall. Mater. Trans.* **29A** 2011
[26] Komura S, Furukawa M, Horita Z, Nemoto M and Langdon T G 2001 *Mater. Sci. Eng.* **A297** 111
[27] Furukawa M, Horita Z and Langdon T G 2001 *Mater. Sci. Eng.* in press
[28] Iwahashi Y, Horita Z, Nemoto M and Langdon T G 1998 *Acta Mater.* **46** 3317
[28] Iwahashi Y, Horita Z, Nemoto M and Langdon T G 1998 *Metall. Mater. Trans.* **29A** 2503
[30] Furukawa M, Iwahashi Y, Horita Z, Nemoto M, Tsenev N K, Valiev R Z and Langdon T G 1997 *Acta Mater.* **45** 4751
[31] Hasegawa H, Komura S, Utsunomiya A, Horita Z, Furukawa M, Nemoto M and Langdon T G 1999 *Mater. Sci. Eng. A* **265** 188
[32] Berbon P B, Komura S, Utsunomiya A, Horita Z, Furukawa M, Nemoto M and Langdon T G 1999 *Mater. Trans. JIM* **49** 772
[33] Komura S, Horita Z, Furukawa M, Nemoto M and Langdon T G 2001 *Metall. Mater. Trans.* **32A** 707
[34] Stolyarov V V, Zhu Y T, Lowe T C and Valiev R Z 2001 *Mater. Sci. Eng. A* **303** 82
[35] Akamatsu H, Fujinami T, Horita Z and Langdon T G 2001 *Scripta Mater.* **44** 759
[36] Nakashima K, Horita Z, Nemoto M and Langdon T G 2000 *Mater. Sci. Eng. A* **281** 82
[37] Nishida Y, Arima H, Kim J-C and Ando T 2000 *J. Japan Inst. Light Metals* **50** 655
[38] Nishida Y, Arima H, Kim J-C and Ando T 2000 *J. Japan Inst. Metals* **64** 1224

Chapter 13

Metal–ceramic nanocomposites

De Liang Zhang

Introduction

This chapter reviews metal–ceramic nanocomposites, while chapters 9 and 14 describe other kinds of nanocomposite materials. The thinking behind the design of a few distinguishable groups of metal–ceramic nanocomposites are analysed, including the well known oxide dispersion strengthened (ODS) alloy nanocomposites, metal nanoparticles embedded in a ceramic matrix with unique optical properties, and ceramics toughened by nanometre sized metallic phases. Different processes which have been used in producing metal–ceramic nanocomposites are reviewed. The state of the art of understanding mechanical and physical properties of these materials and their future applications are discussed. Overall, the majority of the research work in this exciting area is still on materials processing and preparation techniques.

By definition, metal–ceramic nanocomposites are metal–ceramic composite materials with at least one dimension of either the metal phase or ceramic phase smaller than 100 nm. There are several types of metal–ceramic nanocomposites: nanometre sized ceramic particles dispersed in a metal matrix; nanometre sized metal (or semiconductor) wires in a ceramic matrix; nanometre scaled multilayer metal–ceramic composites, and so on. Figure 13.1 illustrates the three types of metal–ceramic nanocomposites categorized by the form of the nanosized phase as particles, wires or thin layers. Similar to metal–ceramic composites with a larger structure scale, metal–ceramic nanocomposites combine the dramatically different natures of metals or semiconductors and ceramics, and may offer unique or enhanced mechanical, physical and chemical properties. The unique or improved properties originate from the very small size of the metal or ceramic phase domains and/or the very large area of metal/ceramic interfaces. Interestingly, the challenge of preparing the materials also originates from these two attributes. The opportunity for creating new commercially attractive materials with highly desirable and often tailored properties has attracted

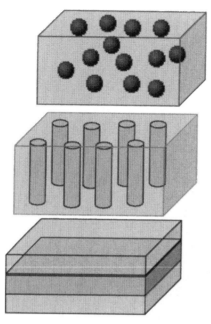

Figure 13.1. Schematic illustration of three major types of metal–ceramic nanocomposite: (a) nanoparticles embedded in a matrix; (b) nanowires embedded in a matrix; and (c) nanometre sized multilayer thin films.

strong interest from a wide range of researchers. This strong interest is reflected by a substantial number of research papers (in the order of one to two hundreds) on this topic published in mainstream English language materials journals and major conference proceedings in the past 4–5 years.

Materials design

The aim of developing a new material is to achieve unique or enhanced mechanical, physical or chemical properties through creating a particular microstructure. This means that the development of a new material always has an element of materials design. For metal–ceramic nanocomposites, the relevant materials design is based on well-known and fundamental aspects of materials science and engineering. Different considerations are relevant to different types of composite.

Oxide dispersion strengthened alloy nanocomposites

Oxide dispersion strengthened alloys have been developed based on the understanding that the strength of metals, especially high temperature strength, and their creep resistance can be enhanced through ceramic dispersion strengthening. This design intent has become fairly well known and is widely used in developing ceramic dispersion strengthened nickel-based superalloys, aluminium alloys, magnesium alloys and even NiAl intermetallic compounds [1–6]. The common feature of this type of material is that ceramic particles with a diameter less than 100 nm are dispersed in a metal matrix. This materials design is based on the understanding that hard ceramic particles strengthen the metal matrix through creating Orowan loops. A simple model which describes the strengthening effect of the particles is illustrated by [7]

$$\tau = \tau_s + 2T/bL \tag{13.1}$$

and

$$L = (6V/\pi)^{-1/3} d_m \tag{13.2}$$

where τ is the yield strength of the composite in shear, τ_s is the threshold stress for moving the dislocations in the matrix, T is the line tension of the dislocations, b is the Burgers vector, L is the mean interparticle spacing, V is the volume fraction of the particles, and d_m is the mean particle diameter. From equations (13.1) and (13.2), it is clear that decreasing the size of the particles is far more effective in increasing the strength than increasing the volume fraction of the ceramic reinforcement. On the other hand, there is a lower limit of particle size at which the strength ceases to increase with decreasing particle size, as shown schematically in figure 13.2 [8]. When the particles are too small, they can be easily cut or climbed over by the dislocations. Once this occurs, equations (13.1) and (13.2) become invalid. The critical particle diameter is in the order of a few nanometres for widely used ceramic particles such as Y_2O_3, Al_2O_3 and SiC.

Dispersions of intermetallic precipitates provide the major strengthening mechanism in precipitation hardened alloys. In comparison, ceramic dispersions are not as effective in strengthening alloys at room temperature. However, ceramic particles are more stable than intermetallic precipitates at

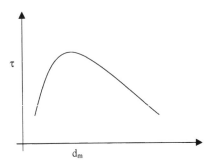

Figure 13.2. Schematic diagram showing the shear yield strength of an oxide dispersion strengthened alloy nanocomposite as a function of the mean particle size of the ceramic reinforcement.

elevated temperatures, so their strengthening effect remains largely unchanged as the temperature increases. This enables oxide dispersion strengthened alloy nanocomposites to exhibit better high temperature strength and creep resistance than precipitation hardened alloys. A dispersion of nanometre sized ceramic particles also has a Zener drag effect on grain growth of the metal matrix, again promoting to higher stability for the microstructure.

Metal nanoparticles or nanowires embedded in a ceramic matrix

It has been found that, because of the quantum confinement effect, some metal or semiconductor nanoparticles (e.g. Pt, Ag and Si) with a diameter of a few nanometres have unique optical properties (e.g. [9, 10]). In order for the nanoparticles to be used to make useful materials, they need to be embedded in a functionally inactive (i.e. inert) matrix such as SiO_2 or a functionally active matrix (i.e. one which interacts physically with the particles) such as TiO_2 and MgO [9]. The major function of the matrix is to provide a solid medium to disperse and support the nanoparticles. There are also cases where silicon wires with nanometre sized diameter are grown in a ceramic such as alumina as a matrix, as shown in figure 13.3 [11].

Metal nanoparticles strengthened ceramics

The design principle of this group of materials is that a dispersion of metal particles with a diameter smaller than 100 nm can increase the bending

Figure 13.3. Transmission electron micrograph showing uniform diameter silicon nanowires (the bright phase) embedded in an alumina matrix [11].

strength of ceramic materials [e.g. 12]. This is attributed to the beneficial effect of the metal nanoparticles in enhancing the sinterability of the ceramic powder and in increasing the stress required to nucleate cracks from flaws. Interestingly, it has been observed [12, 13] that the nanoparticles have little effect on the fracture toughness of the ceramic materials. In order for the metal phase to be effective in increasing the fracture toughness of the ceramic matrix, the size of the metal particles should be at submicrometre or micrometre level.

Soft magnetic nanoparticles in a hard magnetic matrix

The metal–ceramic nanocomposite concept has also been used in developing superhard magnetic materials. It has been well established that by combining a nanocrystalline hard magnet matrix such as $Nd_2Fe_{14}B$ and soft magnet nanoparticles such as α-Fe to form a metal–ceramic nanocomposite, the size of the magnetic hysteresis loop, and thus the energy product of the nanocomposite magnet, can be significantly increased (e.g. [14, 15]). Chapter 20 provides an overview on these nanocomposite permanent magnets.

Processing

The majority of the research on metal–ceramic nanocomposites is about processing of the materials. Researchers are still working on ways of preparing the nanocomposite materials and the relationship between processing conditions and microstructure. Several different techniques have been used to prepare metal–ceramic nanocomposites, including:

- co-sputtering,
- co-evaporation,
- powder blending or high energy mechanical milling plus powder consolidation,
- partial reduction of a ceramic or partial oxidation of a metal,
- decomposition and devitrification,
- ion implantation.

Co-sputtering

The co-sputtering technique is used to prepare metal–ceramic nanocomposite thin films, especially metal nanoparticles embedded in a ceramic matrix such as Si/SiO_2 and Pt/TiO_2 nanocomposites [9, 10], or metal nanoparticles in a semiconductor matrix such as Ag/Si nanocomposite [16]. With co-sputtering, metal and ceramic targets are used at the same time in the sputtering process. The volume fraction of the metal nanocomposites can be adjusted by adjusting the area fraction of the metal target in relation to the ceramic target. The size of the nanoparticles is controlled by a sputtering condition such as deposition rate and substrate temperature.

Co-evaporation and gas condensation

The co-evaporation technique has been used by Yamamoto *et al* [17] to make superparamagnetic Fe_2O_3/Ag nanocomposites. Co-evaporation together with gas condensation has been used to make Ag-Fe alloys with a microstructure of Fe nanoparticles embedded in a Ag matrix, followed by oxidization to convert Fe into Fe_2O_3. Co-evaporation has also been used to produce BiSb-SiO_2 nanocomposites [18].

Powder blending or high energy mechanical milling plus powder consolidation

Powder blending or high energy mechanical milling plus powder consolidation has been widely used in preparing metal–ceramic nanocomposites (e.g. [2, 6, 14, 19–22]). When powder blending plus powder consolidation is used, the process requires one of the starting powders to consist of nanoparticles, i.e. be a nanopowder, since there is no milling process to reduce the particle size of the constituent phases. Metal or ceramic nanopowders can be produced using mechanochemical processing processes [23] or laser-induced gas phase reaction [2, 5, 15]. In blending a nanopowder and other powders, it is important to ensure complete break-up of nanoparticle agglomerates. Because of the large surface area and high total surface energy associated with the nanoparticles, agglomeration is very common, and it is difficult to break up the agglomerates and disperse the nanoparticles homogeneously in a ceramic or metal matrix [2, 5]. Figure 13.4 shows the microstructure of a Mg–3 vol% SiC nanocomposite produced using powder blending [5]. The inhomogeneity of the SiC nanoparticle distribution is clearly visible. In order to ensure a full dispersion of the nanoparticles in the matrix, high energy mechanical milling may have to be used. Powder blending has also been used to produce metal-carbon nanotube composites [24, 25].

High energy mechanical milling is different from simple mechanical powder blending, even though it also often starts with a mixture of metal and ceramic powders. With high energy mechanical milling, plastic deformation, fracturing and cold welding under an inert atmosphere allows destruction of the starting powder particles and creation of new composite powder particles, as shown schematically in figure 13.5. In the initial stage,

Figure 13.4. Mg–3 vol% SiC nanocomposite produced by simple blending [5].

the microstructure of the composite powder particles is at the micrometre level. With further milling, the metal phase is deformed and fractured, while the ceramic phase is mainly fractured. For a metal matrix composite, the ceramic particles are continually fractured into smaller particles, so the ceramic particle size keeps decreasing until such a point that the particle fracture strength is equal to or greater than the stress caused by collisions during milling. Often this balance point corresponds to a ceramic particle size of less than 100 nm.

As an example, figure 13.6 shows the development of the powder particle microstructure of a Cu–20 wt% W soft metal–hard metal nanocomposite [26]. In the very initial stage of milling, only a few W particles are incorporated into each of the Cu particles. Then with further milling which

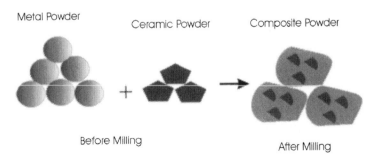

Figure 13.5. Schematic diagram showing the formation of composite powder after high energy mechanical milling.

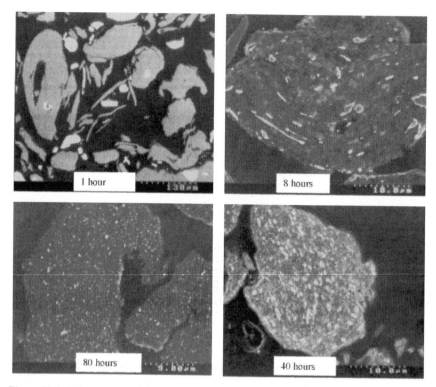

Figure 13.6. Microstructural evolution of powder particles during high energy mechanical milling of Cu–20 wt% W powder. The dark phase is Cu and the bright phase is W. (Courtesy of Jennie Richmond.)

causes working hardening of the Cu and W phases, most of the W particles are deformed, fractured and incorporated into the Cu matrix, forming Cu/W composite powder particles as shown in figure 13.6. With continued milling, the W particles are reduced to nanometre sizes through continued fracturing. When the W particles reach a critical size, further fracturing becomes impossible during milling, and the composite microstructure stops changing. The critical W particle size has not yet been determined. When the W particles are even smaller, perhaps <5 nm in diameter, it is possible that the Cu/W surface energy is high enough to overcome the huge thermodynamic barrier to alloying Cu and W. The W particles then lose their stability and dissolve into the Cu matrix, forming a Cu(W) solid solution or a Cu–W amorphous phase. Attempts to produce nonequilibrium Cu–W amorphous alloy through mechanical alloying have been made by Gaffet *et al* [27] and Zhang and Massalski [28]. The amorphous phase seems to form after very lengthy milling, but the findings are not yet conclusive due to possible oxygen contamination.

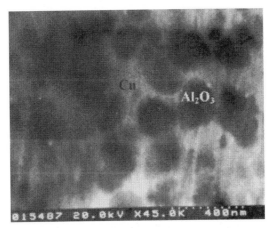

Figure 13.7. SEM micrograph showing the microstructure of a Cu–20 vol% Al$_2$O$_3$ nanocomposite produced by mechanical milling and sintering. (Courtesy of Danyang Ying.)

Metal–ceramic nanocomposites with nanometre sized ceramic particulate reinforcement or ceramic–metal nanocomposites with nanometre sized metallic particles can also be produced by a similar technique [29]. Since the nanostructure of each powder particle evolves through numerous deformation, fracturing and cold welding events during milling, a homogeneous powder is achieved at the same time as the nanostructure. Sometimes the as-milled nanocomposite powder is not stable, and is converted into a different metal–ceramic nanocomposite structure during subsequent heat treatment or sintering. One example of this case is the production of a Cu-Al$_2$O$_3$ nanocomposite powder from a Cu(Al)-CuO nanocomposite powder produced by high energy mechanical milling of a Cu(Al)-CuO powder mixture, as shown in figure 13.7 [19]. This technique has also been used to produce Al(Cu)-Al$_2$O$_3$ and other M-Al$_2$O$_3$ nanocomposites, where M is a metal or a metal–aluminium solid solution produced by reducing a metal oxide with aluminium [30, 31].

Following powder blending or high energy mechanical milling, the nanostructured powder can be consolidated using conventional powder metallurgy. In consolidating the powder, it is essential to maintain the nanostructure. For some metal–ceramic systems, such as Cu/Al$_2$O$_3$, it is not difficult to maintain the nanostructure during consolidation, since the sintering temperature is far below the melting point of the ceramic phase. For other metal–ceramic systems, such as Ni/Al$_2$O$_3$, it is difficult to maintain the nanostructure during pressureless sintering. In these cases, a pressure based process such as hot isostatic pressing is often then employed to lower the sintering temperature down to a level well below the melting point of the ceramic phase.

Mechanical alloying is also sometimes used to produce metal–ceramic nanocomposites directly from elemental powders [32, 33].

Partial reduction of a ceramic or partial oxidation of a metal

When partial reduction of a ceramic is used to produce a metal–ceramic nanocomposite, a nanostructured ceramic powder is first produced by using a chemical route such as precipitation [14, 34–36]. Then the ceramic powder is partially reduced, normally using a mixture of argon and hydrogen, to convert one of the ceramic phases into metal nanoparticles. Normally this ceramic phase is an oxide of a relatively inactive metal such as Cu, W or Ni. The powders are then consolidated using conventional sintering or hot pressing. The partial reduction technique is mainly used to produce ceramic matrix nanocomposite powders.

When partial oxidation of metal is used to produce a metal–ceramic nanocomposite, an alloy powder is first produced using atomization or mechanical alloying. The alloy powder is then partially oxidized to bring the active elements such as Al out of the alloy to form oxide particles. The formation of oxide particles involves nucleation and growth in the solid state, and the process condition can thus be easily tailored to obtain oxide nanoparticles dispersed in a metal matrix. The partial oxidation technique is usually more suitable for manufacturing composites with a small volume fraction (less than 5%) of oxide nanoparticles.

Decomposition and devitrification

Metal–ceramic nanocomposites can be produced by decomposing a metastable oxide such as SiO_x ($x < 2$) [37] or FeO [38], or by crystallizing an amorphous alloy such as $Nd_2(Fe,Co,Nb)_{15}B$ (39) or Ni-Ta-C [40]. During controlled decomposition or crystallization, metallic, semiconductor or ceramic nanoparticles can be produced and are embedded in a metal or ceramic matrix.

Ion implantation

McHargue *et al* [41] have used ion implantation to produce single crystal alumina (sapphire)-Fe nanocomposites. In this process, [56]Fe ions are injected into the ceramic material and then the material is annealed at an elevated temperature to precipitate Fe nanoparticles in the alumina matrix, as shown in figure 13.8. A similar technique has also been used by Johnson and co-workers [42, 43] to produce metal–metal nanocomposites such as Pb or Pb-Sn nanoparticles embedded in an Al matrix, and Pb-Cd nanoparticles embedded in a Si matrix.

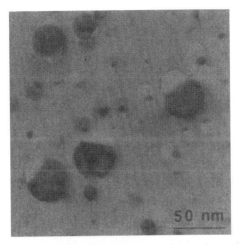

Figure 13.8. Transmission electron micrograph of α-alumina implanted with Fe ions after being annealed for 1 h at 1500 °C [41]. The dark particles are Fe.

Properties

The study of microstructure/property relationships in metal–ceramic nano-composites is still at an early stage. Many papers dealing with processing of metal–ceramic nanocomposites have also reported the results of testing of properties such as strength, creep resistance [2, 6] and photoluminoscenece intensity [9, 10]. However, there is still a lack of systematic studies on the effects of microstructural parameters such as particle size and volume fraction on nanocomposite properties. Among the different groups of metal–ceramic nanocomposites, the mechanical properties, especially the creep resistance, of oxide dispersion strengthened alloy nanocomposites have been well studied [2, 6, 8], and are perhaps best understood. The magnetic properties of some of the nanocomposite magnetic materials have also been well studied [e.g. 44–46). In 1995 and 1996, Holtz and Provenzano published two papers [47, 48] with an intention to study systematically the mechanical properties of metal–ceramic nanocomposites, using model soft metal–hard metal systems. Unfortunately, the results of the study were compromised by oxidation during processing, so the conclusions are not very convincing.

In relation to the properties of metal–ceramic nanocomposites, there are a few critical questions to be answered, including:

What is the critical size of ceramic (or hard metal) particles to have a significant reinforcement effect? Establishing the strength–particle size relationship of a metal ceramic nanocomposite is essential for defining this critical size. A similar question arises for the toughening effect of nanometre sized metallic particles in ceramic matrix composites.

How does the strength and ductility of a metal–ceramic nanocomposite change with the volume fraction of the ceramic nanoparticles? So far research has shown that a few volume percent of ceramic nanoparticles can cause a significant increase in high temperature strength and creep resistance; room temperature strength and ductility have not been studied. The strength increase with increasing volume fraction of nanoparticles is not expected to be linear, so it is important to determine the critical point after which the strength increase diminishes. For precipitation hardening alloys such as Al-Cu alloys, 5 vol% of precipitates with a size of 5–20 nm can increase the yield strength of aluminium from approximately 10 MPa to 300–400 MPa [49]. Such a significant strengthening effect has not been reported for metal–ceramic nanocomposites.

How do metal–ceramic nanocomposites fracture? For metal matrix composites with micrometre sized ceramic reinforcement particles, fracture is often initiated by cracking of the ceramic particles under high levels of stress concentration. However, metal–ceramic nanocomposites might exhibit very different fracture behaviour. It is very difficult to open a crack in a ceramic particle of <100 nm in size, because of high particle strength and the lack of defects. In addition, stress concentrations around the nanoparticles may not be very high given that the nanoparticles can relieve the stress by rotation or dislocation climb. The strength of the matrix/particle interface plays a critical role in defining the fracture behaviour of the nanocomposites. With low levels of stress concentration and difficulties in cracking the nanoparticles, deformation of the nano-composites is expected to be homogeneous, and may lead to improved ductility [49].

What are the physical and chemical properties of metal–ceramic nanocomposites? There are few data concerning the physical and chemical properties of metal–ceramic nanocomposites, except for the popular hard magnet–soft magnet nanocomposites and a few nanocomposites developed as optical materials.

Microstructural stability

Like any nanostructured or nanosized materials, the microstructural stability of metal–ceramic nanocomposites is a critical issue. The metal/ceramic interface energy and the mobility of atoms at the interface play a critical role in determining the microstructural stability during processing and service at elevated temperatures. In general, metal/ceramic interfaces make a positive contribution to the Gibbs energy of the metal/ceramic system, so they increase the driving force for the reaction between the metal and ceramic phases. For metal–ceramic nanoparticles, this contribution is substantial, and in many cases could be overwhelming due to the large area of the interfaces. It could make normally inert metal–ceramic couples (e.g. Cu and Al_2O_3) reactive. The high mobility of atoms at the metal/ceramic interfaces can also greatly enhance reaction kinetics and make reactions occur at much lower temperatures. Recently, in studying the processing of Ti_3Al/SiC nanocomposites, Liang *et al* [29] found that SiC nanoparticles cannot survive hot isostatic pressing of the composite powder at 800 °C, while micrometre sized SiC particles in the same material remain stable during pressing at this temperature.

There has been much research into the microstructural stability of nanocrystalline alloys, but research into the microstructural stability of metal–ceramic nanocomposites is still very limited. However, published research on melting of nanoparticles embedded in a solid may be helpful in understanding the microstructural stability of metal–ceramic nanocomposites (e.g. [50–53]).

Applications

Unlike monolithic alloys, ceramics or polymers, the beauty of the metal–ceramic composites is that they offer much greater flexibility for tailoring the microstructure to develop the properties needed for particular applications. This means that the development of metal–ceramic nanocomposites will be more and more targeted to particular applications. There have already been several examples of successful applications of metal–ceramic nanocomposites, including oxide dispersion strengthened alloy or intermetallic nanocomposites and nanocomposite hard permanent magnets. Recently a Japanese steel maker, NKK, has developed a steel-based metal–ceramic nanocomposite which demonstrates high strength (yield strength = 780 MPa) and much improved ductility and formability [54]. This steel-based nanocomposite is to be used for making automotive underbody parts such as suspensions.

Concluding remarks

Research on metal–ceramic nanocomposites is very active in the international materials science and engineering community due to the potential commercial opportunities and the challenge that this class of materials can offer. Up to now, it seems that the majority of the research published still deals with the art of processing the nanocomposite materials. Due to this effort, processing condition/microstructure relationships in metal–ceramic nanocomposites have been reasonably well understood, and several processing techniques have been established. However, the microstructural stability of the nanocomposites and the metal–ceramic reactions during processing and services at elevated temperatures are still critical issues which have not been very well addressed.

The amount of published work on microstructure/property relationships in metal–ceramic nanocomposite is still small, so there is a lack of in-depth knowledge in this area. The materials design intent is in general clear for each group of metal–ceramic nanocomposites, and is normally based on the fundamental principles of materials science and engineering. The creep resistance of ceramic nanoparticle-reinforced metals or alloys has been extensively studied. It has been clearly demonstrated that the ceramic nanoparticles have a significant positive effect on creep resistance. In a similar manner, the magnetic properties of nanocomposite magnets which combine ceramic hard magnetic phases and metallic soft magnetic nanoparticles are promising. The optical properties of metal (mainly Pt and Ag) and semiconductor (mainly Si and CdS) nanoparticles embedded in an inactive or active ceramic matrix are also highly attractive.

There has been no reported research 'outside of the square' in the area of metal–ceramic nanocomposites. The properties of some metal–ceramic nanocomposites with almost equal volume fractions of nanometre sized metal and ceramic phase particles might possess unique and perhaps very important mechanical, physical or chemical properties.

References

[1] Benjamin J S 1970 *Met. Trans.* **1** 2943
[2] Ma Z Y, Li Y L, Liang Y, Zheng F, Bi J and S C Tjong 1996 *Mater. Sci. Eng. A* **219** 229
[3] Ma Z Y, Tjong S C, Li Y L and Liang Y 1997 *Mater. Sci. Eng. A* **225** 125
[4] Trojanova Z, Lukae P, Ferkef H, Mordike B L and Riehemann W 1997 *Mater. Sci. Eng. A* **134–236** 798
[5] Ferkel H and Mordike B L 2001 *Mater. Sci. Eng. A* **298** 193
[6] Arzt E and Grahle P 1998 *Acta Mater.* **46** 2717
[7] Arsenault R J 1984 *Mater. Sci. Eng.* **64** 171
[8] Arzt E 1998 *Acta Mater.* **46** 5611
[9] Sasaki T, Koshizaki N, Terauchi S, Umehara H, Matsumoto Y and Koinuma M 1997 *Nanostruct. Mater.* **8** 1077
[10] Koshizaki N, Umehara H, Sasaki T and Oyama T 1997 *Nanostruct. Mater.* **8** 1085
[11] Ong P P, Zhu Y and Wang H 2000 *Proc. 1st Int. Conf. on Advanced Materials Processing* ed. D L Zhang, K L Pickering and X Y Xiong (Melbourne: Institute of Materials Engineering Australasia) p 391
[12] Niihara K and Suzuki Y 1999 *Mater. Sci. Eng. A* **261** 6
[13] O Sbaizero and G Pezzotti 2000 *Acta Mater.* **48** 985
[14] Kim T H, Yu J H and Lee J S 1997 *Nanostruct. Mater.* **9** 213
[15] Liu H, Wang L, Wang A, Lou T, Ding B and Hu Z 1997 *Nanostruct. Mater.* **9** 225
[16] Niu F, Chang I T H, Dobson P J and Cantor B 1997 *Mater. Sci. Eng. A* **226–228** 161
[17] Yamamoto T A, Shull R D and Hahn H W 1997 *Nanostruct. Mater.* **9** 539

[18] Brochin F, Devaux X, Ghanbaja J and Scherrer H 1999 *Nanostruct. Mater.* **11** 1

[19] Ying D Y and Zhang D L 2001 *Mater. Sci. Eng. A* **301** 90

[20] Goodwinn T J, Yoo S H, Matteazzi P and Groza J R 1997 *Nanostruct. Mater.* **8** 559

[21] Matteazzi O and Alcala M 1997 *Mater. Sci. Eng. A* **230** 161

[22] Jose J and Khadar M A 2001 *Acta Mater.* **49** 729

[23] Ding J, Miao W F, McCormick P G and Street R 1995 *Phys. Lett.* **67** 380

[24] Dong S R, Tu J P and Zhang X B 2001 *Mater. Sci. Eng. A* **313** 83

[25] Flahaut E, Peigney A, Laurent Ch, Marliere Ch, Chastel F and Rousset A 2000 *Acta Mater.* **48** 3803

[26] Zhang D L, Ying D Y and Richmond J J 1999 *Proc. NZ–Korea Seminar on Engineering Materials, University of Waikato, Hamilton, New Zealand*, February, p 102

[27] Gaffet E, Louison C, Harmelin M and Faudot F 1991 *Mater. Sci. Eng. A* **134** 1380

[28] Zhang D L and Masslaki T B 1997 *Advanced Materials and Development* ed. W G Ferguson and W Gao (University of Auckland, NZ) p 387

[29] Liang J, Zhang D L, Li Z W and Gao W unpublished research

[30] Wu J M and Li Z Z 2000 *J. Alloys Compounds* **299** 9

[31] Osso D, Le Caer G, Begin-Colin S, Mocellin A and Matteazzi P 1993 *J. de Phys. IV, Colloque C7* **3** 1407

[32] Zhou L Z, Guo J T and Fan G J 1998 *Mater. Sci. Eng. A* **249** 103

[33] Koch C C 1998 *Mater. Sci. Eng. A* **244** 39

[34] Carles C, Brieu M and Rousset A 1997 *Nanostruct. Mater.* **8** 529

[35] Hyuga H, Hayyashi Y, Sekino T and Niihara K 1997 *Nanostruct. Mater.* **9** 547

[36] Oh S, Sando M, Sekino T and Niihara K 1998 *Nanostruct. Mater.* **10** 267

[37] Kahler U and Hofmeister H J 2000 *Metastable Nanocryst. Mater.* **8** 488

[38] Tokumitsu K and Nasu T 2000 *J. Metastable Nanocryst. Mater.* **8** 562

[39] Chen Z, Zhang Y, Hadjipanayis G C, Chen Q and Ma B 1999 *J. Alloys Compounds* **287** 227

[40] Wilde J R and Greer A L 2001 *Mater. Sci. Eng. A* **304–306** 932

[41] McHargue C J, Ren S X and Hunn J D 1998 *Mater. Sci. Eng. A* **253** 1

[42] Johnson E, Johansen A, Sarholt L and Dahmen U 2001 *J. Metastable Nanocryst. Mater.* **10** 267

[43] Sarholt L, Jensen A S B, Touboltsev V S, Johansen A and Johnson E 2001 *J. Metastable Nanocryst. Mater.* **10** 283

[44] Panagiotopoulos I and Hadjipanayis G C 1998 *Nanostruct. Mater.* **10** 1013

[45] Chen Z, Zhang Y and Hadjipanayis G C 1999 *Nanostruct. Mater.* **11** 1285

[46] Zhang H W, Zhang W Y, Yan A R, Sun Z G, Shen B G, Tung I C and Chin T S 2001 *Mater. Sci. Eng. A* **304–306** 997

[47] Holtz R L and Provenzano V 1996 *Nanostruct. Mater.* **7** 905

[48] Provenzano V and Holtz R L 1995 *Mater. Sci. Eng. A* **304** 125

[49] Polmear I 1995 in *Light Alloys: Metallurgy of Light Metals* 3rd edition (London: Arnold)

[50] Zhang D L and Cantor B 1991 *Acta Metall. Mater.* **39** 1595

[51] Zhang D L, Hutchinson J and Cantor B 1994 *J. Mater. Sci.* **29** 2147

[52] Lu K and Jin Z H 2001 *Curr. Opin. Solid State Mater. Sci.* **5** 39

[53] Sheng H W, Lu K and Ma E 1998 *Acta Mater.* **46** 5195

[54] *Materials World* July 2001 p 9

Chapter 14

Alumina/silicon carbide nanocomposites

Richard Todd

Introduction

Alumina/silicon carbide nanocomposites typically consist of an alumina matrix with a conventional grain size of 2–5 μm, containing a dispersion of silicon carbide nanoparticles of ∼200 nm diameter (figure 14.1). Interest was first aroused in these materials with the publication in 1991 of a paper by Niihara [1] reviewing work on a number of ceramic composites containing nanoscale phases. The alumina/silicon carbide system stood out as showing dramatic improvements in strength and maximum useful temperature compared with unreinforced alumina. The reported improvement in room temperature strength of nearly 300% was particularly impressive, and this could be improved further, to more than 400%, simply by annealing in air or inert gas. These results have stimulated much research around the world, and the conclusions of these endeavours were initially controversial. A clearer picture of the behaviour of alumina/silicon carbide nanocomposites is now beginning to emerge, however, and whilst it seems that the large increases in strength reported initially were due mainly to microstructural refinements that can be obtained more easily by conventional means, there is a real nanocomposite effect in the response of these materials to grinding and wear which is genuinely novel. The property improvements in this respect are as dramatic as the strength improvements originally reported, and are on the verge of commercial exploitation. Further opportunities exist in the high temperature mechanical properties of these materials, which also show significant improvements over alumina. This chapter summarizes the current knowledge of alumina matrix nanocomposites, with particular emphasis on the alumina/silicon carbide system which has received most attention. Chapters 9 and 13 describe other kinds of nanocomposite materials.

Processing and microstructural development

Most alumina-based nanocomposites have been made by conventional processing of mixed, commercial powders of alumina and silicon carbide. These are readily available with the fine particle sizes required. The highly refractory, covalent silicon carbide particles hinder densification significantly owing to their reluctance to participate in diffusional processes, and the majority of research to date has employed hot pressing to obtain full density. Pressureless sintering to high density is possible for low silicon carbide contents ($\leq$5%), however, either by raising the temperature to >1700 °C [2] or by using sintering aids such as yttria [3] to reduce the temperature required. Yttria has a detrimental effect on the sintering of pure alumina, and recent work [4] has suggested that its beneficial effect in the nanocomposites owes much to the presence of silica on the surface of the silicon carbide particles. This may react with the yttria and alumina to form a thin layer of liquid phase at the grain and phase boundaries at sintering temperatures, thus avoiding the alumina/silicon carbide interfaces which are thought to contribute to the difficulty in sintering nanocomposites in the absence of additives. Pressureless sintered materials retain the commercially important improvements in wear resistance and polishing behaviour that are seen in hot pressed materials. The fact that these property improvements can be obtained with only 2% silicon carbide gives further scope for lowering the sintering temperature.

The powders used for the matrix and nanoparticles are usually of similar initial size, typically 200 nm. During densification, the alumina grains grow until the pinning effect of the silicon carbide particles prevents further growth. If the grains are allowed to grow to stagnation, the final grain size agrees reasonably well with the Zener pinning model [5]. The silicon carbide particles do not coarsen and are immobile for the same reason that they hinder densification. The alumina grain growth therefore envelops some of the silicon carbide particles so that they are within the grains in the final microstructure, with others remaining on the grain boundary (figure 14.1). It has been widely observed that larger particles tend to be on the grain boundary, whilst smaller particles are intragranular, and this has often been attributed to the greater pinning force of an individual particle the larger its size. There is a geometrical reason for this observation as well, however, in that for a given position of its centre, a particle is more likely to impinge on a grain boundary if its radius is bigger. Careful analysis of experimental results shows that whilst the interaction between particles and grain boundary does increase the number of intergranular particles, this is true for all particle sizes, and the greater proportion of large particles on the grain boundary is mainly due to the geometrical effect described above.

Other processing schemes have been used to obtain different microstructures. Winn and Todd [6] used a coarse alumina powder (3.5 µm) to prevent

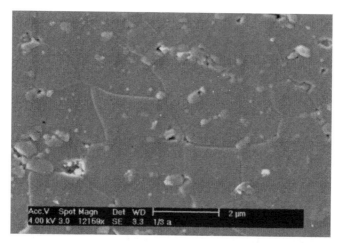

Figure 14.1. Scanning electron micrograph of typical nanocomposite microstructure, showing 2–3 μm alumina grains containing a dispersion of SiC nanoparticles. The particles range in size from 20 to 500 nm, and can be found both on the alumina grain boundaries and within the grains. A SiC agglomerate can be seen towards the top of the picture.

grain growth from leading to intragranular particles, so that a predominantly intergranular microstructure was produced (figure 14.2). Boehmite sols have also been used to produce the alumina matrix [7–9], and this has generally led to a better dispersion of silicon carbide particles and a smaller alumina grain size. Chemical processing routes have also been used for the silicon carbide. There are several reports (e.g. [10]) of the use of polymeric precursors such as

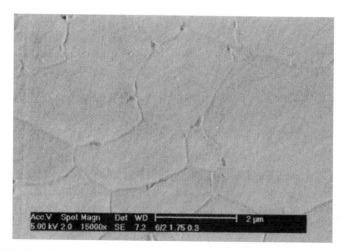

Figure 14.2. Scanning electron micrograph of 'intergranular' nanocomposite microstructure, showing that the majority of the SiC particles are located on the alumina grain boundaries.

polysilastyrene, both in conjunction with alumina powder, and with a Boehmite precursor for the matrix, as above. The resulting silicon carbide particles are much finer (typically 10–100 nm) than those available in powder form and the silicon carbide agglomerates that are always present in powder-sourced material (figure 14.1) are absent. It is thought that the absence of agglomerates is responsible for the high strengths shown by these materials compared with powder processed material.

Room temperature strength and toughness

Niihara's original paper [1] reported an increase in strength from 350 MPa for pure alumina to 1050 MPa on the addition of 5% silicon carbide, and a further increase to 1520 MPa on annealing at 1300 °C for 1 h in air or inert gas. The accompanying increase in toughness from 3.5 to 4.8 MPa m$^{1/2}$ was more modest, 37%, and the conclusion was therefore that the strength increase was largely a consequence of a reduction in the critical flaw size.

The reported strength increase on adding silicon carbide nanoparticles to alumina has received much attention since Niihara's publication [1] and, despite many attempts to reproduce this result, none has been successful. Most workers have obtained some increase in strength, but the extent of the improvement ranges from 0% [11] to 50% [12] compared with alumina of comparable grain size, with most showing an increase of only ∼20%. The discrepancy between Niihara's result and those of others has been the subject of much debate, but careful examination of the evidence reveals a simple story which can explain the disagreement. The key is to observe that part of the reason for the very high strength increase of the nanocomposite relative to the alumina reported by Niihara is the low strength of the alumina used to make the comparison (350 MPa). This suggests that the alumina used suffered from a microstructural defect such as abnormal grain growth, and indeed Niihara comments in [1] that the silicon carbide additions prevented the formation of abnormal grains. The alumina used by most other workers did not suffer from abnormal grain growth and therefore had higher strengths of 500 MPa or more, which would have reduced substantially the fractional increase in strength calculated for the nanocomposite. The fact that Niihara's nanocomposite had somewhat higher strengths than those of other workers may be attributable to the fact that the quoted strengths are for small specimens in 3-point bending, whereas other studies have used 4-point bend tests, which give lower strength values, and also to the normal differences in processing between different laboratories. In summary, the majority of the large strength increase reported by Niihara seems to be the result of the suppression of abnormal grain growth by the silicon carbide additions. Other workers did not observe this because they used aluminas for comparison with the nanocomposite which were already free of abnormal grain growth. Given that abnormal grain growth can be avoided by cheaper and easier methods than the addition of silicon carbide (e.g. the addition of a trace of magnesium oxide, MgO), the effect is of little commercial interest.

The modest strength increase of ~20% found by most workers when comparing materials with the same grain size is similarly of little commercial value. The small extent of the improvement also makes its origin difficult to identify. Careful measurements of toughness [13, 14] have shown no evidence for an increase in the nanocomposites relative to alumina, so other effects must be responsible. Among those suggested are a reduction in critical

Figure 14.3. Fracture surfaces of alumina (top) and an alumina–11 vol% SiC nanocomposite, showing change in fracture mode from intergranular in the alumina to largely transgranular in the nanocomposite.

flaw size owing, for example, to enhanced milling properties of the composite powder [15], or increased resistance to microcrack initiation because of the steeper R-curve during initial crack growth, which may be expected from the transgranular fracture mode [16]. Another factor which must be involved is the improvement in surface finish resulting from the silicon carbide additions, as described later in this chapter. This may lead to smaller surface cracks and increased levels of compressive stress in the surface following grinding or polishing [17].

There may be some truth in all of these suggestions, and it is important to note that it is not surprising that the strengths of the nanocomposites and pure alumina are not identical, because their fracture behaviour also shows a striking contrast. Whilst alumina is well known to exhibit almost exclusively intergranular fracture, the fracture mode of the nanocomposites is almost exclusively transgranular, as shown in figure 14.3, even for volume fractions of silicon carbide as low as 1% [6]. In contrast to the strength improvements, the transgranular fracture mode of the nanocomposites has proved to be reproducible, and all workers agree on this for standard material containing ~5% silicon carbide with a particle size of 200–300 nm. Nonetheless, the explanation for this apparent strengthening of the grain boundaries has also been a source of debate, although most suggestions rely in some way on the large stresses which arise from the thermal expansion mismatch between the silicon carbide particles and the alumina matrix during cooling from sintering temperatures. Alumina has a bigger thermal expansion coefficient than silicon carbide and during cooling the matrix shrinks around the particles putting them into compression. The thermal stresses are large, approaching −2000 MPa in the particles [18], as shown in figure 14.4, and can therefore be expected to influence crack propagation. The volume averaged stresses measured by neutron diffraction shown in figure

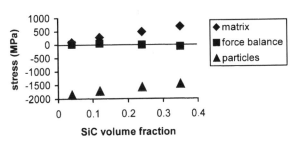

Figure 14.4. Volume-averaged hydrostatic thermal stress in the alumina matrix and SiC particles as a function of volume fraction for alumina–260 nm SiC nanocomposites, measured by neutron diffraction. The 'force balance' is the sum of the stresses in the matrix and the particles weighted by their volume fraction, and should be zero in the absence of an externally applied stress. The fact that this is so to within experimental error validates the results (the error bars are smaller than the symbols).

14.4 are consistent with the predictions of an elastic model which can be used to estimate the spatial variation of the stresses. The model correctly predicts that the volume averaged matrix stress is tensile on average, but close to the particles the radial stress is predicted to be compressive and the hoop stress is tensile.

Niihara originally suggested that tensile thermal hoop stresses in the matrix around silicon carbide particles situated within the alumina grains could attract the crack out of the grain boundary and into the body of the grain, causing transgranular fracture. Levin *et al* [19] have also suggested that the compressive radial stresses around the particles act to strengthen the grain boundaries. We have used the results shown in figure 14.4 with the elastic model which these results support to perform a fracture mechanics analysis to test these hypotheses. The results show that although particles within the grains do indeed provide a driving force for cracks to leave the grain boundary, the effect is far too weak to be effective, as shown in figure 14.5(a). The compressive radial stress around a typical silicon carbide particle situated on the grain boundary, however, causes a local increase of about 50% in the apparent toughness of the boundary, as shown in figure 14.5(b). This is sufficient to bring the local Mode I fracture toughness up to the level of the grain interior, so that cracks running along grain boundaries inclined to the plane of maximum tensile stress will tend to be deflected into the grain to

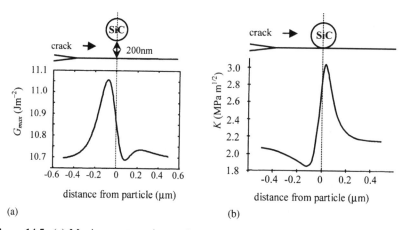

(a)   (b)

Figure 14.5. (a) Maximum attractive strain energy release rate into the grain body as a crack passes an intragranular particle surrounded by thermal stresses. Transgranular fracture will occur if G_{max} rises from the grain boundary value by about a factor of two, to $24\,\mathrm{J\,m^{-2}}$. The peak value falls far short of this, however, being only about 4% higher than the grain boundary value. (b) The apparent toughness of the grain boundary as a grain boundary crack passes a particle on the grain boundary. The apparent toughness increases by about 50% owing to the compressive radial stresses around the SiC particle, taking it close to the matrix value of $3.1\,\mathrm{MPa\,m^{1/2}}$.

become a transgranular crack. Local strengthening of the grain boundary by intergranular particles also results from the fact that the particle/matrix interface is thought to be tougher than the grain boundary [20], and by the high toughness and stiffness of the silicon carbide particles themselves. Direct observation of crack paths has confirmed that it is the silicon carbide particles situated on the grain boundaries which are directly responsible for the change in fracture mode [21].

The further increase in strength on annealing reported by Niihara [1] has been reproduced by several researchers, although the extremely high reported strength, 1520 Mpa, has not been reproduced. The effect results primarily from crack healing in the nanocomposites, which is facilitated by (i) the formation of silica by oxidation of the silicon carbide particles, which fills in cracks, (ii) a greater retention of surface compressive stress than in alumina, owing to the superior creep resistance of the nanocomposites (see below), which helps surface crack healing and causes a direct strengthening effect, and (iii) the flat intergranular crack faces, which can fit back together easily during healing, unlike the tortuous and interlocked crack faces of the alumina [22, 23]. The extent of the improvement obtained depends on the annealing atmosphere, with air being better than (nominally) inert gas [24], presumably due to factor (i).

High temperature mechanical properties

The mechanical properties of alumina/silicon carbide nanocomposites are also improved at high temperature. The ductile–brittle transition temperature is increased, and the hardness is increased above the transition temperature. This has been attributed to dislocation pinning by the silicon carbide particles [1]. The nanocomposites retain their room temperature strength to higher temperatures ($>1000\,°C$) than alumina ($800\,°C$) [1], presumably due in part to the annealing effect described above. The creep resistance is also improved dramatically, with the creep rate being reduced by two to four orders of magnitude compared with alumina, depending on the silicon carbide content, stress and loading method [25], and the creep life is increased by typically one order of magnitude. This is because the silicon carbide particles inhibit grain boundary sliding and diffusion creep, owing again to their refractory nature and covalent bonding. The particles also cause cavitation, however, as a result of which the failure strain is reduced relative to that of alumina.

These improvements in high temperature strength suggest a potential market for alumina/silicon carbide nanocomposites, as they provide a material which is intermediate in both performance and price between pure alumina on the one hand, and silicon carbide, SiAlONs and Si_3N_4 on the other.

Grinding and wear behaviour

It was noticed at an early stage that alumina/silicon carbide nanocomposites are much easier to polish metallographically than alumina [22], because they do not suffer from the propensity for grain pullout by grain boundary fracture that plagues alumina. The nanocomposites are also more resistant to environments involving severe wear, including erosive wear [26, 27], abrasive wear [6] and sliding wear [28, 29], in which the transition from mild wear in the early stages to severe wear involving surface fracture is either delayed or is entirely absent. The resistance to mild wear, however, is not improved [29].

The reductions in wear rate for a given treatment and alumina grain size are significant, typically a factor of 2–3, and reproducible. Improvements of this level can be obtained with as little as 2% silicon carbide. Figure 14.6 shows surfaces of alumina and a nanocomposite following grinding with 45 μm diamond paste [6]. There is a marked difference between the behaviour of alumina and the nanocomposites. The worn surface of the alumina shows evidence of extensive intergranular fracture and grain pullout. The nanocomposite, however, shows much less surface fracture, and the dominant material removal mechanism appears to be plastic ploughing and cutting of the surface by the abrasive particles. Although some surface fracture is always present in the nanocomposites, the individual pullouts are smaller than in pure alumina of the same grain size and account for less of the total area, and the fracture mode within the pullouts is mainly transgranular.

The reduction in surface fracture and pullout is responsible for the improved wear resistance. This is demonstrated in figure 14.7 [30], which shows a plot of the wear rate in a standard load-controlled abrasive wear test against the area fraction of the worn surface suffering from fracture and grain pullout, both for a large number of alumina matrix nanocomposites containing various particle compositions, sizes and volume fractions, and with various grain sizes, and for pure aluminas, also with a range of

Figure 14.6. Surfaces of alumina (left) and alumina–11 vol% SiC nanocomposite of similar grain size after grinding with 45 μm diamond particles.

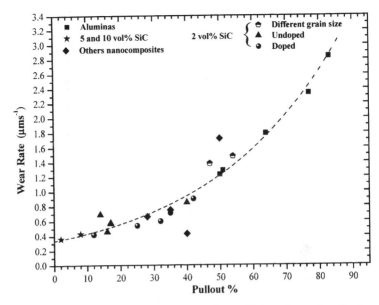

Figure 14.7. Wear rate versus percentage of surface pulled out by surface fracture for a wide range of nanocomposites and pure aluminas.

grain sizes. All the materials fall on the same line, implying that the improved wear resistance is attributable solely to the greater resistance to grain pullout of the nanocomposites. Furthermore, it shows that in other aspects of the wear mechanism, the nanocomposites behave the same as the aluminas. Ortiz Merino and Todd [31] have shown that the increase in wear rate as the area fraction of pullout increases can be explained quantitatively using a simple model for the acceleration of material removal from the smooth, unfractured parts of the surfaces by plastic ploughing and cutting. The larger the area fraction of pullout, the greater the force per unit area on the remaining load bearing area, and the higher the rate at which this surface is ground away.

There is as yet no unequivocal answer to the crucial question of what causes the striking reduction in surface fracture and pullout in the nanocomposites. The apparent strengthening of the grain boundaries and change in fracture mode are clearly related to the change in mode observed in conventional fracture surfaces shown in figure 14.3. The change in fracture mode represents only a local strengthening of the grain boundaries in the vicinity of the silicon carbide particles, however, and it is difficult to see how this can provide an effect sufficiently potent to explain the results directly, especially with volume fractions of silicon carbide as low as 2%. Davidge *et al*'s suggestion [16] that the change in fracture mode produces a more steeply rising R-curve for very short cracks in the nanocomposites may be relevant in this context, but it should be borne in mind that this effect, which arises from the greater toughness of the grains compared with the grain boundaries, cannot be very great, otherwise the fracture mode would remain intergranular.

An additional, purely geometrical explanation has been put forward recently [31] which stems from the fact that the size of the individual surface pullouts is observed to be significantly smaller in the nanocomposites than in alumina. This may be related to the change in fracture mode in the nanocomposites. A pullout in alumina, in which the fracture is intergranular, has to be at least a significant fraction of a grain in size, whereas the transgranular fracture in the nanocomposites allows much smaller pieces of material to be removed. This is important to wear in general because the amount of material removed for each pullout is proportional to the pullout volume. In the case of grinding and polishing of surfaces, the area fraction of pullout at steady state is also roughly proportional to the volume of the pullouts, because as well occupying a larger area of the surface, larger pullouts are deeper and therefore take longer to remove by continued polishing.

These ideas based on a change in fracture mode caused by a local strengthening of the grain boundaries at intergranular particles are consistent with the main features of grinding and polishing in alumina and alumina matrix nanocomposites. Great improvements in grinding and polishing can be produced by low volume fractions of silicon carbide because only a few particles per grain facet are required to deflect the crack into the grain interior. The intergranular nanocomposite shown in figure 14.2 exhibits the same improvements in surface finish because it is the grain boundary particles that are thought to be responsible for the change in fracture mode. The importance of having small silicon carbide particles is that if the particles are too large, they become too far apart, and large pullouts and extensive intergranular fracture can occur between them. Finally, it is found that silicon carbide nanoparticles are more effective than other particles that have been investigated, such as titanium nitride TiN and silicon nitride Si_3N_4, because they have the best combination of higher toughness and stiffness than alumina coupled with a lower thermal expansion coefficient, all factors that favour the local grain boundary strengthening thought to be responsible for the change in fracture mode.

Summary and outlook for exploitation

The mechanical and tribological properties of alumina/silicon carbide nanocomposites have been thoroughly characterized over the past decade. It has become clear that the improvements in strength and toughness that they offer are modest, and are probably not commercially viable given the difficulty of making the materials compared with alumina. The high temperature properties, polishing and wear behaviour offer very considerable improvements over monolithic alumina, however, to such an extent that the increase in expense of these materials may be cost-effective. This is particularly true in applications where the nanocomposites can challenge materials such as silicon carbide and silicon nitride, which are themselves more expensive to make than the nanocomposites. Applications can be anticipated in components for use at high temperatures or in highly abrasive or erosive environments, and in components such as ceramic seals which need high quality surface finishes.

Academically, the explanations for the property improvements are becoming clearer. It seems that a few effects underpin most of the property improvements. These are as follows:

1. The inability of the silicon carbide particles to participate in diffusion processes. This hinders sintering, but also reduces the grain size and limits grain boundary sliding at high temperature.
2. The change in fracture mode from intergranular in alumina to transgranular in the nanocomposite. This appears to be a consequence of local deflection of cracks from the boundary into the grain at intergranular particles. The particles can deflect cracks by virtue of the compressive residual stresses surrounding them, their high toughness and stiffness, and their strong interface with the matrix.
3. The change in fracture mode reduces the volume of material removed in individual grain pullout events, which leads in turn to slower wear rates, better surface finishes, and consequently improved strength.
4. The change in fracture mode may also lead to an initially steeper R-curve, which inhibits crack initiation, contributing to the wear resistance and strength increase.

References

[1] Niihara K 1991 *J. Ceram. Soc. Jpn.* **99** 9742
[2] Stearns L C, Zhao J and Harmer M P 1992 *J. Eur. Ceram. Soc.* **10** 473
[3] Jeong Y K and Niihara K 1997 *NanoStruct. Mater.* **9** 193;
    Jeong Y K, Nakahira A and Niihara K 1999 *J. Am. Ceram. Soc.* **82** 3609
[4] Cock A M, Todd R I and Roberts S G unpublished research
[5] Borsa C E, Jiao S, Todd R I and Brook R J 1995 *J. Microscopy* **177** 305
[6] Winn A J and Todd R I 1999 *Br. Ceram. Trans.* **98** 219
[7] Xu Y, Nakahira A and Niihara K 1994 *J. Ceram. Soc. Jpn.* **102** 312
[8] Borsa C E, Jones N M R, Brook R J and Todd R I 1997 *J. Eur. Ceram. Soc.* **17** 865
[9] Winn A J, Wang Z C, Todd R I and Sale F R 1998 *Silicates Industriels* **63** 147
[10] Sternitzke M, Derby B and Brook R J 1998 *J. Am. Ceram. Soc.* **81** 41
[11] Poorteman M, Descamps P, Cambier F, O'Sullivan D, Thierry B and Leriche A 1994 *Proc. 8th CIMTEC, Firenze*

[12] Carroll L, Sternitzke M and Derby B 1996 *Acta Mater.* **44** 4543

[13] Meschke F, Alves R P and Schneider G A 1997 *J. Mater. Res.* **12** 3307

[14] Hoffman M and Rödel J 1997 *J. Ceram. Soc. Jpn.* **105** 1086

[15] Sternitzke M 1997 *J. Eur. Ceram. Soc.* **17** 1061

[16] Davidge R W, Brook R J, Cambier F, Poorteman M, Leriche A, O'Sullivan D, Hampshire S and Kennedy T 1997 *Br. Ceram. Trans.* **96** 121

[17] Wu H Z, Roberts S G and Derby B 2001 *Acta Mater.* 49 507

[18] Todd R I, Bourke M A M, Borsa C E and Brook R J 1997 *Acta Mater.* **45** 1791

[19] Levin I, Kaplan W D, Brandon D G and Layyous A A 1995 *J. Am. Ceram. Soc.* **78** 254

[20] Jiao S and Jenkins M L 1998 *Phil. Mag. A* **78** 507

[21] Jiao S, Jenkins M L and Davidge R W 1996 *J. Microscopy* **185** 259

[22] Zhao J, Stearns L C, Harmer M P, Chan H M and Miller G A 1993 *J. Am. Ceram. Soc.* **76** 503

[23] Wu H Z, Lawrence C W, Roberts S G and Derby B 1998 *Acta Mater.* **46** 3839

[24] Chou I A, Chan H M and Harmer M P 1998 *J. Am. Ceram. Soc.* **81** 1203

[25] Ohji T, Nakahira A, Hirano T and Niihara K 1994 *J. Am. Ceram. Soc.* **77** 3259

[26] Walker C N, Borsa C E, Todd R I, Davidge R W and Brook R J 1994 *Br. Ceramic Proc.* **53** 249

[27] Twigg P C, Alexander G, Davidge R W and Riley F R 1997 *Key Eng. Mater.* **132–136** 1520

[28] Rodríguez J, Martín A, Pastor J Y, Llorca J, Bartolomé J and Moya J 1999 *J. Am. Ceram. Soc.* **82** 2252

[29] Chen H J, Rainforth W N and Lee W E 2000 *Scripta Mater.* **42** 555

[30] Ortiz Merino J L 2001 DPhil Thesis, University of Oxford

[31] Ortiz Merino J L and Todd R I unpublished research

SECTION 3

MAGNETIC NANOMATERIALS

One of the most important potential and actual applications of nanocrystalline materials is in magnetic devices and components. Nanocrystalline materials exhibit outstanding magnetic properties, which are very desirable in many industrial and domestic applications. Chapter 15 describes the fundamental magnetic behaviour of nanograined materials; chapters 16 and 17 describe the magnetic properties of a variety of different nanocrystalline alloys; chapters 18 and 19 describe industrial applications of soft ferromagnetic nanocrystalline alloys; and chapter 20 describes industrial applications of hard ferromagnetic nanocrystalline alloys.

Chapter 15

Microfabricated granular films

K Takanashi, S Mitani, K Yakushiji and H Fujimori

Introduction

Insulating granular films consisting of magnetic metal particles such as cobalt embedded in an insulating matrix such as alumina exhibit large tunnel magnetoresistance. The large charging energy of particles is expected to lead to the appearance of single electron tunnelling phenomena. This chapter describes spin-dependent single electron tunnelling in microfabricated granular systems. Two types of device structures have been fabricated: granular nanobridge and current-perpendicular-to-plane geometry structures. The granular nanobridge structure consists of electrodes separated by a nanometre-sized gap in which a thin Co-Al-O granular film is filled. Coulomb blockade with a clear threshold voltage V_{th} is observed at 4.2 K, and tunnel magnetoresistance is enhanced at a voltage slightly above V_{th}. This enhancement is explained by the orthodox theory of single electron tunnelling in ferromagnetic multiple junctions. The current-perpendicular-to-plane structure consists of a thin Co-Al-O granular film sandwiched by top and bottom electrodes with a submicron-sized contact area. Clear Coulomb staircases are observed by the addition of an Al-O bottleneck layer that leads to an asymmetric configuration of tunnel resistances. Furthermore, tunnel magnetoresistance shows oscillatory behaviour associated with the Coulomb staircase. This chapter demonstrates how the voltage control of tunnel magnetoresistance can be realized using single electron tunnelling phenomena, which will be useful for application to tunnel magnetoresistance devices such as magnetic random access memories.

Background

Tunnel magnetoresistance has attracted much attention in recent years because of potential applications for magnetic heads, magnetic random

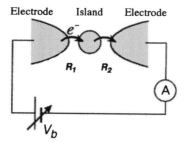

Figure 15.1. Schematic illustration of double tunnel junctions, basic for observing single electron tunnelling.

access memories, and other spin-electronic devices [1]. Most studies to date have been devoted to magnetic tunnel junctions with trilayer ferromagnetic metal/insulator/ferromagnetic metal structures, with junction areas the order of $0.01\,\mu m^2$ at the smallest. A large tunnel magnetoresistance ratio reaching approximately 50% has been achieved at room temperature [2].

Recent development of nanofabrication techniques enables the formation of different types of magnetic nanostructures leading to the emergence of novel magnetotransport phenomena. Suppose that the size of a magnetic tunnel junction gets smaller and smaller. What would then happen to tunnel magnetoresistance? The reduction in junction size gives rise to an increase in the charging energy, and single electron tunnelling phenomena are expected to appear if the charging energy overcomes thermal fluctuations. Figure 15.1 shows a schematic illustration of double tunnel junctions, which is a basic structure to observe single electron tunnelling. This structure consists of small island electrodes that are separated by tunnel barriers. The tunnelling of electrons can be inhibited and the current does not flow at small bias voltages if the electrostatic energy $e^2/2C$ of a *single* excess electron on the island is much larger than the thermal energy $k_B T$, where C is the capacitance of the island. The suppression of current at small bias voltages is called the Coulomb blockade. When the bias voltage increases and exceeds a threshold $V_{th} = e/2C$, the current starts to increase. If the junction resistances of two barriers are similar ($R_1 \approx R_2$), the current increases smoothly with the bias voltage, as shown in figure 15.2(a). On the other hand, if the difference between two junction resistances is very large ($R_1 \ll R_2$ or $R_1 \gg R_2$), the current increases stepwise with bias voltage depending on the number of electrons accumulating on the island, as shown in figure 15.2(b). The step-like structure in the current–bias voltage (I–V_b) characteristic is called the Coulomb staircase. The Coulomb blockade and the Coulomb staircase are representative phenomena of single electron tunnelling. In practice, even for multiple junctions including more than one island between the electrodes, single electron tunnelling phenomena can be observed.

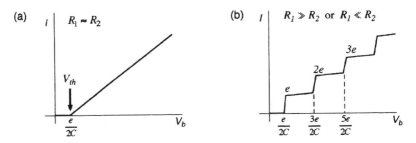

Figure 15.2. Current–bias voltage (I–V_b) characteristics for single electron tunnelling in double junctions: (a) with no asymmetry of tunnnel resistances and (b) with strong asymmetry of tunnel resistances.

For the appearance of single electron tunnelling phenomena, the following two conditions are generally essential. Firstly, as mentioned above, the charging energy E_c required to add an electron to an island must far exceed the thermal energy, i.e.

$$E_c \gg k_B T. \tag{15.1}$$

Secondly a junction resistance R_J must exceed the resistance quantum $R_Q = h/e^2$, i.e.

$$R_J \gg R_Q \tag{15.2}$$

which ensures that the wave function of an excess electron on an island is localized there. Without this condition, the electron would be delocalized through the island, permitting transport even for a bias voltage lower than the threshold voltage V_{th}. Single electron tunnelling phenomena in *nonmagnetic* systems, where both islands and electrodes are nonmagnetic, have already been extensively investigated theoretically and experimentally [3].

In *magnetic* systems, the interplay of spin-dependent tunnelling and single electron tunnelling is expected to give rise to remarkable tunnel magnetoresistance phenomena. It has been predicted theoretically by several groups [4–11] that the enhancement and oscillation of tunnel magnetoresistance seems to be due to spin-dependent single electron tunnelling. However, there have been only a few experiments [12–16] to date on single electron tunnelling in magnetic systems, since it is difficult to fabricate appropriate sample structures to observe spin-dependent single electron tunnelling. The islands should be small enough to satisfy equation (15.1), since the charging energy E_c increases with decreasing the island size. The charging energy of an isolated island is described as

$$E_c = e^2 / 4\pi\varepsilon d \tag{15.3}$$

where d is the diameter assuming the island is a sphere, and ε is the dielectric constant of an insulator around the island. Although the charging energy E_c is somewhat modified by the configuration of surrounding islands and electrodes in a real system, we may consider that the charging energy is roughly proportional to the inverse of the island diameter $1/d$.

A possible method for preparing small magnetic islands is the use of a microfabrication technique such as electron beam lithography. However, the microfabrication technique is usually limited to the formation of submicron-sized islands. E_c for submicron-sized islands is generally of the order of 10^{-4}–10^{-3} eV (1–10 K). Therefore, single electron tunnelling phenomena can be observed only at very low temperatures [12]. Granular films have an advantage because nanometre-sized particles are naturally formed by self-assembling processes, and single electron tunnelling phenomena may appear at higher temperatures. This chapter reviews recent studies of spin-dependent single electron tunnelling in granular films containing magnetic particles.

Insulating granular films

Insulating granular films consist of small metallic particles embedded in an insulating matrix. If the metal content is lower than the percolation limit, transport is dominated by tunnelling of electrons between particles. If the particles are magnetic, tunnel magnetoresistance appears because the random magnetization vectors on particles at zero or low applied fields become aligned as the applied field is increased, leading to a decrease in resistivity. The pioneering works on tunnel magnetoresistance in insulating granular films were done in Ni-Si-O films by Gittleman *et al* [17] and by Helman and Abeles [18]. However, the magnitudes of tunnel magneto-resistance were very small. In 1995 a large tunnel magnetoresistance was reported in Co-Al-O granular films by Fujimori *et al* [19] Figure 15.3 shows a typical example of the applied field dependence of tunnel magneto-resistance in a Co-Al-O granular film. A large tunnel magnetoresistance ratio

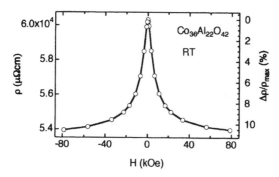

Figure 15.3. Applied field dependence of tunnel magnetoresistance at room temperature for a $Co_{36}Al_{22}O_{42}$ film.

(a)   (b)

Figure 15.4. (a) Plan view and (b) cross-sectional transmission electron micrographs for a $Co_{46}Al_{19}O_{35}$ film.

of 10% is obtained even at room temperature. Tunnel magnetoresistance has been found to increase with decreasing temperature, and exceed 20% at low temperatures [20]. The atomic composition of the film shown in figure 15.3 is $Co_{36}Al_{22}O_{42}$, as determined by Rutherford backscattering spectroscopy.

Insulating granular films are usually prepared by sputtering and evaporation techniques. Reactive sputtering is particularly useful for Co-Al-O granular films [21]. A cobalt–aluminum alloy target is sputtered in a mixture of argon and oxygen. The aluminum is then selectively oxidized, and a granular structure is formed of nanometre-sized cobalt metal particles in an insulating aluminum oxide matrix. Figure 15.4 shows plan view and cross-sectional transmission electron micrographs for a $Co_{46}Al_{19}O_{35}$ film [21]. The film has an isotropic granular structure consisting of cobalt particles of 2–3 nm in diameter (dark spheres) and intergranular aluminum oxide about 1 nm in thickness (white channels). Figure 15.5 shows a high-resolution transmission electron micrograph for a $Co_{52}Al_{20}O_{28}$ film, indicating that a crystalline cobalt particle is surrounded by non-stoichiometric aluminum oxide with an amorphous structure [21]. Co-Al-O granular films are superparamagnetic at room temperature. The size distributions of the cobalt particles can be investigated by analysing their superparamagnetic behaviour, and results are consistent with the transmission electron micrograph observations [22].

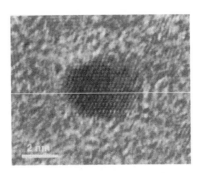

Figure 15.5. High-resolution transmission electron micrograph for a $Co_{52}Al_{20}O_{28}$ granular film.

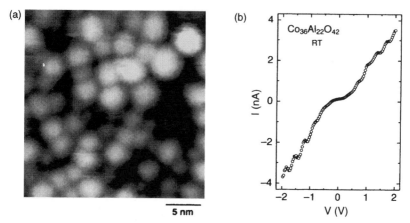

Figure 15.6. (a) Scanning tunnelling microscopy topographic image and (b) current–bias voltage I–V_b curve at room temperature for a $Co_{36}Al_{22}O_{42}$ film.

The particle sizes in insulating granular films are so small that single electron tunnelling is expected to occur between each particle. In a granular film with a macroscopic size containing a large number of particles, however, single electron tunnelling phenomena are averaged out due to the large distributions of particle sizes and interparticle distances [13, 23]. The tunnel paths should be restricted to observe clearly the single electron tunnelling phenomena. A simple method to restrict the tunnel paths is to use scanning tunnelling microscopy. The tunnel path on the surface is limited to only one particle just below the scanning tunnelling microscope tip. Clear Coulomb staircases are observed in the I–V_b measurements for Co-Al-O granular films even at room temperature [24, 25]. Figures 15.6(a) and (b) show typical examples of a scanning tunnelling microscope topographic image and an I–V_b curve respectively, for a $Co_{36}Al_{22}O_{42}$ film.

A more advantageous method for a variety of measurements and applications than scanning tunnelling microscopy is to fabricate a device structure consisting of a small part of a granular film with microscopic electrodes. This chapter discusses two types of device structure in Co-Al-O granular films manufactured by a focused ion beam etching technique. One type of device structure is granular nanobridge structure [26], in which point-shaped electrodes are separated by a very narrow lateral gap, filled by a Co-Al-O granular film. The other type of device structure is a current-perpendicular-to-plane structure [27], in which a thin Co-Al-O granular film is sandwiched by ferromagnetic electrodes and the current flows in the perpendicular direction to the film plane through a few cobalt particles. The current–bias voltage curves show enhancement and oscillation of tunnel magnetoresistance due to spin-dependent single electron tunnelling.

Granular nanobridge

Experimental results

A schematic view of a granular nanobridge structure is shown in figure 15.7. The granular nanobridge has been fabricated on a glass substrate using the following process: A 15 nm thick NbZrSi amorphous film is deposited by r.f. sputtering, and is formed into source and drain electrodes by focused ion beam etching using 30 kV gallium ions. Figure 15.8 shows a scanning ion microscopy image of electrodes separated by a gap of length $l = 30$ nm and width $w = 60$ nm. Deep trenches (60 nm wide and 200 nm deep) are formed beside the gap by focused ion beam etching to avoid the formation

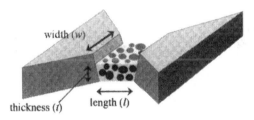

Figure 15.7. Schematic view of a granular nanobridge structure.

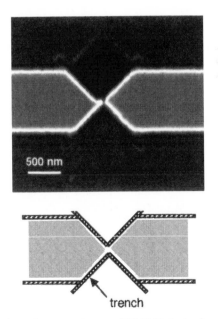

Figure 15.8. Scanning ion microscopy image of NbZrSi electrodes separated by a nano-metre-sized lateral gap. The length l and width w of the gap are 30 and 60 nm respectively.

of unnecessary current paths outside the gap. A $Co_{36}Al_{22}O_{42}$ granular film of thickness t is deposited on the patterned surface by reactive r.f. sputtering, and the gap is filled by the $Co_{36}Al_{22}O_{42}$ film. The aspect ratio of the trenches is so high that they are not filled with the $Co_{36}Al_{22}O_{42}$ film. The characteristic sizes of a granular nanobridge, i.e. the width, length and thickness (w, l and t), can be varied over the ranges 60–700 nm, 30–70 nm and 5–30 nm, respectively. Current–bias voltage curves are measured at 4.2 K using an electrometer with a two-terminal arrangement. The tunnel magnetoresistance is defined as $\Delta R/R_{H=0}$, i.e. the resistance change divided by the resistance at zero applied field, and is evaluated from the difference between the current–bias voltage curves curves at applied fields of $H = 0$ and 10 kOe.

Figure 15.9(a) shows current–bias voltage curves at magnetic field levels of $H = 0$ (solid lines) and $H = 10\,kOe$ (dashed lines) for a sample with $w = 60\,nm$, $l = 30\,nm$ and $t = 7.5\,nm$. Here, the threshold voltage V_{th} is

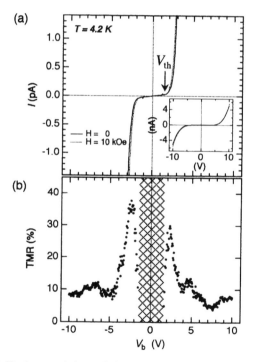

Figure 15.9. (a) I–V_b characteristics and (b) V_b dependence of tunnel magnetoresistance measured at 4.2 K for a granular nanobridge with $w = 60\,nm$, $l = 30\,nm$ and $t = 7.5\,nm$. In (a), the solid and dashed curves represent I–V_b curves at $H = 0$ and 10 kOe respectively. I–V_b curves throughout the measured current range are shown in the inset. V_{th} is the threshold voltage ($\sim 1.5\,V$) due to the Coulomb blockade. The values of tunnel magnetoresistance shown in (b) have been evaluated from the solid and dashed curves shown in (a). The hatched area in (b) represents the Coulomb blockade region ($|V_b| < V_{th}$).

defined as that below which the current is zero within the accuracy of 100 fA, and it is approximately 1.5 V in this case. In the range of $|V_b| < V_{th}$, a Coulomb blockade occurs, and the current increases rapidly when $|V_b|$ exceeds V_{th}. A Coulomb blockade has not clearly been observed in Co-Al-O granular films of macroscopic sizes, because macroscopic samples contain a broad distribution of cobalt particle sizes, and the tunnelling of electrons between large particles with small E_c can start at small voltages. In a granular nanobridge, however, the tunnel paths are so limited that the Coulomb blockade is very strong.

Figure 15.9(b) shows the bias voltage dependence of tunnel magneto-resistance. Tunnel magnetoresistance depends strongly on bias voltage. For $|V_b| < 4.0$ V, tunnel magnetoresistance increases with decreasing $|V_b|$ and reaches a maximum value larger than 30% at the voltage slightly above the threshold voltage V_{th} (~ 1.5 V). In the Coulomb blockade region, i.e. $|V_b| < V_{th}$ (hatched area), there is little quantitative reliability on the measurements because the current is very low (< 100 fA). For $|V_b| > 4.0$ V, on the other hand, tunnel magnetoresistance is about 8%, showing no large change with bias voltage, although there is still weak oscillatory behaviour in the bias voltage dependence of tunnel magnetoresistance.

Similar results have been obtained in other granular nanobridges with different sizes. Figures 15.10(a) and (b) show the current–bias voltage curve and the bias voltage dependence of tunnel magnetoresistance respectively, for a sample with $w = 700$ nm, $l = 40$ nm and $t = 15$ nm. The threshold voltage V_{th} is observed to be 0.4 V, which is lower than that in the sample shown in figure 15.9. The threshold voltage shows a tendency to increase with a decrease in the size of the granular nanobridge. The voltage V_p, where the tunnel magnetoresistance shows a maximum, is slightly larger than the threshold voltage V_{th}. Figure 15.11 shows V_p versus V_{th} in granular nanobridges of different sizes. A clear correlation between them is seen, suggesting that the enhanced tunnel magnetoresistance is caused by the Coulomb blockade.

Theoretical explanation

The orthodox theory of single electron tunnelling is used to explain the experimental results, particularly the enhanced tunnel magnetoresistance near the threshold voltage V_{th}. In the orthodox theory, the tunnel path is modelled as an equivalent electrical circuit. For a granular nanobridge, a parallel circuit of triple-tunnel-junctions is assumed, as shown in figure 15.12(a). This is the simplest model to explain the experimental results because at least two magnetic particles are needed in each series of junctions to study the spin-dependent transport in nanobridges with nonmagnetic electrodes. The higher-order tunnelling process, called co-tunnelling [9, 13], is neglected, because the tunnel resistances between particles and between

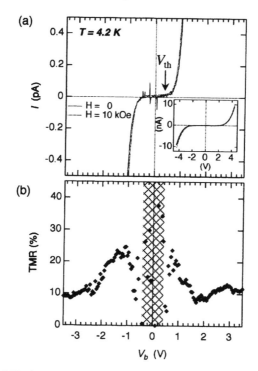

Figure 15.10. (a) I–V_b characteristics and (b) V_b dependence of tunnel magnetoresistance measured at 4.2 K for a granular nanobridge with $w = 700$ nm, $l = 40$ nm and $t = 15$ nm. In (a), the solid and dashed curves represent I–V_b curves at $H = 0$ and 10 kOe respectively. I–V_b curves throughout the measured current range are shown in the inset. The hatched area in (b) represents the Coulomb blockade region ($|V_b| < V_{th}$).

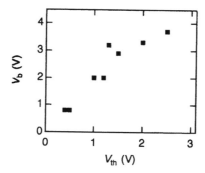

Figure 15.11. V_p versus V_{th} for granular nanobridges with different sizes. V_p is the voltage where the tunnel magnetoresistance shows a maximum, and V_{th} is the threshold voltage in the I–V_b curve.

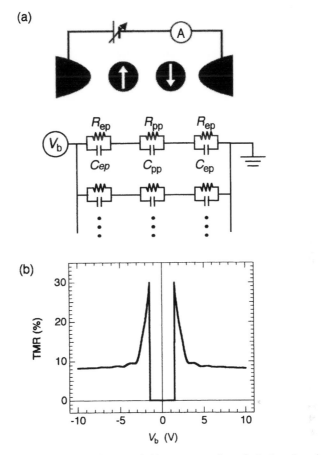

Figure 15.12. (a) Model of a granular nanobridge structure for calculations based on the orthodox theory, with a parallel circuit of 20 triple-tunnel-junctions. (b) Numerical calculations of tunnel magnetoresistance as a function of V_b using the model in (a). For details of the parameters used see the text.

an electrode and a particle are estimated to be about 10^5 times larger than $R_Q \approx 25.8\,\text{k}\Omega$.

In order to obtain the stable tunnelling current, the detailed balance equation for the probability of states $p(\{n_i\}_\alpha)$ is constructed in matrix form:

$$\dot{p} = \mathbf{M}p = 0 \tag{15.4}$$

where $p = (\ldots, p(\{n_i\}_\alpha, \ldots)^{\mathrm{T}}$ and $\mathbf{M}$ is the transition matrix in the configuration space constructed from $\{n_i\}_\alpha$ with the index α labelling the different charge states. The tunnelling current through the kth junction is given by

$$I_k = e \sum_\alpha p(\{n_i\}_\alpha)[\Gamma_k^+(\{n_i\}_\alpha) - \Gamma_k^-(\{n_i\}_\alpha)] \tag{15.5}$$

where $\Gamma_k^{+(-)}(\{n_i\}_\alpha) \propto 1/R_k$ is the forward (backward) tunnelling rate through the kth junction with initial charge state $\{n_i\}_\alpha$. Current conservation requires that the tunnelling current I_k is the same for all the junctions in each series.

The tunnelling current can be evaluated at the junction between a non-magnetic electrode and a particle, where the tunnelling rate $\Gamma_k^{+(-)}(\{n_i\}_\alpha)$ is independent of the magnetic field. The magnetic field dependence of the tunnelling current comes from the probability $p(\{n_i\}_\alpha)$ of the charge state $\{n_i\}_\alpha$ which is determined by equation (15.4). Since the transition matrix **M** contains the tunnelling rates between magnetic particles, the matrix **M** and therefore the probability $p(\{n_i\}_\alpha)$ can be modified by applying the magnetic field.

For a bias voltage V_b just above the threshold V_{th}, a few charge states contribute to the tunnelling current, and tunnelling rates with these charge states are very different from each other due to the charging energy. Therefore, strong modification of the probability $p(\{n_i\}_\alpha)$ is made to satisfy the detailed balance equation and the tunnel magnetoresistance is strongly enhanced just above the threshold. This kind of tunnel magnetoresistance enhancement in double tunnel junctions has been studied by Barnas and Fert [4, 5] and by Majumdar and Hershfield [6]. They also predicted the oscillating behaviour of tunnel magnetoresistance against bias voltage associated with the Coulomb staircase. No Coulomb staircase appears in the current–bias voltage curves of granular nanobridges. However, oscillation of the tunnel magnetoresistance is observed, although the magnitude of the oscillation is small, as shown in figures 15.9 and 15.10. There is no strong asymmetry in tunnel resistances, i.e. no bottleneck in the tunnel paths of granular nanobridges, leading to no Coulomb staircase and only weak oscillation of tunnel magnetoresistance. Moreover, there are many junction arrays in a granular nanobridge and the randomness of junction capacitances also smears out the tunnel magnetoresistance oscillation.

In order to explain the experimental results for a sample with $w = 60\,\mathrm{nm}$, $l = 30\,\mathrm{nm}$ and $t = 7.5\,\mathrm{nm}$, as shown in figure 15.9, consider a parallel circuit of 20 triple-tunnel-junctions, and assume that the tunnel resistance between an electrode and a particle is expressed as $R_{ep} = (1 \pm \delta)\bar{R}_{ep}$, where δ is the deviation from the typical value $\bar{R}_{ep}$. Other junction parameters such as tunnel resistances between particles R_{pp}, junction capacitances C_{ep}, and current perpendicular to the plane are also assumed to be distributed around mean values, i.e. $R_{pp} = (1 \pm \delta)\bar{R}_{pp}$, $C_{ep} = (1 \pm \delta)\bar{C}_{ep}$, and $C_{pp} = (1 \pm \delta)\bar{C}_{pp}$. The deviation δ for each junction parameter is randomly chosen within the range of $-0.1 < \delta < 0.1$. The temperature for single electron tunnelling is $4.2\,\mathrm{K}$ and typical value of tunnel resistances with parallel magnetization vectors are

$$\bar{R}_{pp} = \bar{R}_{ep}/2. \tag{15.6}$$

The tunnel resistance between particles with antiparallel magnetization vectors is larger than between particles with magnetization vectors, and is given by

$$\bar{R}_{pp} = \frac{\bar{R}_{ep}}{2} \frac{1 + P^2}{1 - P^2} \qquad (15.7)$$

where the spin polarization P is assumed to be 0.42 for cobalt. Typical values of junction capacitances are taken to be $\bar{C}_{ep} = 0.1\,\text{aF}$ and $\bar{C}_{pp} = 0.05\,\text{aF}$, which are reasonable considering the average particle size and interparticle distance in Co-Al-O granular films [24, 26].

The bias voltage dependence of tunnel magnetoresistance obtained by this numerical calculation is shown in figure 15.12(b). The theoretical result is in good agreement with experiment. The tunnel magnetoresistance is enhanced just above the threshold voltage $V_{th} \approx 1.5\,\text{V}$, and decreases with bias voltage V_b. The randomness of junction capacitances smears out the oscillation of the total tunnel magnetoresistance as shown in figure 15.12(b). The quantitative difference between threshold and maximum voltages V_{th} and V_p is not essential, but is caused by effects such as current leakage through the glass substrate.

Current perpendicular to plane

The previous section showed how enhanced tunnel magnetoresistance is found in granular nanobridges, where a proper limitation of the number of tunnel paths in granular films makes it possible to observe spin-dependent single electron tunnelling. However, no Coulomb staircase is observed and the oscillation of tunnel magnetoresistance is not clear in granular nanobridges. Strong asymmetry in tunnel resistances leads to the appearance of a Coulomb staircase. In an assembly of particles such as granular films, therefore, the Coulomb staircase is expected to appear when a tunnel resistance between two neighbouring particles or between a particle and an electrode is much larger than the other resistances in the tunnel path. In other words, a bottleneck of tunnel conductance needs to be present somewhere between the electrodes.

In Co-Al-O granular films, the sample geometry of current-perpendicular-to-plane measurements makes it easy to add a bottleneck in the tunnel path, observe a Coulomb staircase and investigate its relationship with tunnel magnetoresistance. Figure 15.13(a) shows a schematic of the current perpendicular to plane geometry, consisting of a Co-Al-O granular film sandwiched by top and bottom ferromagnetic electrodes, with a submicron-sized contact area. A very thin aluminum oxide layer is inserted between the bottom electrode and the Co-Al-O layer as a bottleneck. Samples are prepared on Si/SiO_2 substrates by r.f. sputtering. A 15 nm thick cobalt film is first deposited as a bottom

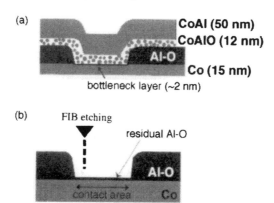

Figure 15.13. Schematic illustrations of (a) a current perpendicular to plane structure and (b) the process of making a contact area by focused ion beam (FIB) etching.

electrode, and then a 40 nm thick aluminum oxide film is deposited on to the cobalt film by using a sintered alumina target. Next, a small contact area is made by focused ion beam etching, as shown in figure 15.13(b). The etching process is carefully performed to leave a very thin aluminum oxide layer on the surface, which contributes to bottleneck formation. After making the contact window, a 1–2 nm thick aluminium oxide layer is deposited. Consequently, the bottleneck is given by the combination of residual and subsequently deposited aluminum oxide. The existence of a bottleneck is important for the appearance of a Coulomb staircase, and samples without a bottleneck show no Coulomb staircase. A 12 nm thick Co-Al-O granular film followed by a 50 nm thick Co-Al top electrode is finally deposited.

Figure 15.14(a) shows current–bias voltage curves at magnetic field levels of $H = 0$ (solid lines) and $H = 10$ kOe (dashed lines) at 4.2 K for a current perpendicular to plane geometry with a $0.5 \times 0.5 \, \mu m^2$ contact area. Clear Coulomb staircases are observed for both current–bias voltage curves. The first three steps from zero bias voltage appear at every 20 mV. However, the steps at higher bias voltage do not keep this regular period. In spite of more than 10^4 parallel tunnel paths between electrodes in a contact area $(0.5 \times 0.5 \, \mu m^2)$ much larger than the cobalt particle sizes (2–3 nm), clear Coulomb staircases appear. This suggests that the current at low bias voltage prefers some restricted local paths that have the lowest charging energy of all the paths available. Similar results were previously reported in the current-perpendicular-to-plane measurements for nonmagnetic granular films [28–30]. It is considered that, at higher bias voltage, the contribution of various tunnel paths including those with higher charging energy leads to the irregular period of the Coulomb staircase.

Figure 15.14(b) shows the bias voltage dependence of tunnel magneto-resistance derived from the two current–bias voltage curves shown in

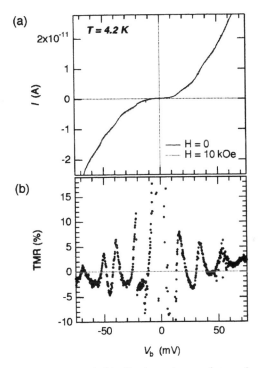

Figure 15.14. (a) $I-V_b$ curves and (b) V_b dependence of tunnel magnetoresistance measured at 4.2 K for a current perpendicular to plane sample with a $0.5 \times 0.5 \, \mu m^2$ contact area. In (a), the solid and dashed curves represent $I-V_b$ curves at $H = 0$ and 10 kOe respectively.

figure 15.14(a). The oscillation of tunnel magnetoresistance is clearly seen as a function of bias voltage with the same period as the Coulomb staircase. Figures 15.15(a) and (b) show current–bias voltage curves and the bias voltage dependence of tunnel magnetoresistance respectively over a magnified range of bias voltage. The modification of the current–bias voltage curve from applying a magnetic field brings about an enhancement of tunnel magnetoresistance at the steps of the Coulomb staircase, resulting in an oscillation in the tunnel magnetoresistance. This is consistent with theoretical calculations in the framework of the orthodox theory [4–8]. The tunnel magnetoresistance shows the largest value at a bias voltage of $V_b = 15 \, mV$, and converges to almost zero with further increase in bias voltage. This is probably because the barrier quality of the aluminum oxide bottleneck layer is still poor, leading to a rapid decrease in tunnel magnetoresistance with increasing bias voltage. The sign of tunnel magneto-resistance changes, and one possible origin of this effect is spin accumulation [4], although further studies are necessary to elucidate this point.

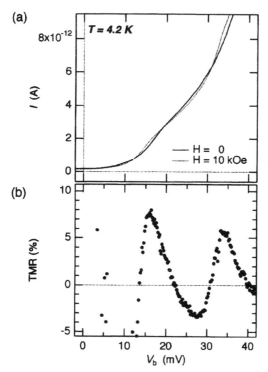

Figure 15.15. (a) $I–V_b$ curves and (b) V_b dependence of tunnel magnetoresistance redrawn over a magnified range of V_b for the same curves as shown in figure 15.14. In (a), the solid and dashed curves represent $I–V_b$ curves at $H = 0$ and 10 kOe respectively.

Summary

Spin-dependent single electron tunnelling phenomena have been investigated in microfabricated granular systems including granular nanobridge and granular current perpendicular to plane structures. Granular nanobridges consisting of electrodes separated by a nanometre-sized gap filled with a Co-Al-O granular film show a Coulomb blockade with a clear threshold voltage (V_{th}) and enhanced tunnel magnetoresistance at a bias voltage V_b slightly above the threshold voltage V_{th}. This enhancement is explained by the orthodox theory of single electron tunnelling in ferromagnetic multiple tunnel junctions. Current perpendicular to plane samples consisting of a thin Co-Al-O granular film sandwiched by top and bottom electrodes in a submicron-sized contact show clear Coulomb staircases and the oscillation of tunnel magnetoresistance associated with the Coulomb staircase. The enhancement and oscillation of tunnel magnetoresistance in microfabricated granular systems are considered to be characteristic phenomena of spin-dependent single electron tunnelling.

This chapter has demonstrated that voltage control of tunnel magneto-resistance can be realized using single electron tunnelling phenomena, which will be useful for application to tunnel magnetoresistance devices such as magnetic random access memory. All measurements have been performed only at 4.2 K. Since the charging energy of particles in granular films is originally larger than the thermal energy of room temperature, however, the appearance of spin-dependent single electron tunnelling is expected even at room temperature, with the following improvements: suppression of leak currents in device structures; and monochromization of particle size distribution, excluding large particles with small charging energy from the tunnel path.

References

[1] For review see Levy P M and Zhang S 1999 *Curr. Opin. Solid State Mater. Sci.* **4** 223

[2] Han X F, Ohgane M, Kubota H ando Y and Miyazaki T 2000 *Appl. Phys. Lett.* **77** 283

[3] For review see Grabert H and Devoret M H (eds) 1992 *Single Charge Tunnelling* NATO ASI Series, vol 294 (New York: Plenum Press)

[4] Barnas J and Fert A 1998 *Europhys. Lett.* **44** 85

[5] Barnas J and Fert A 1998 *Phys. Rev. Lett.* **80** 1058

[6] Majumdar K and Hershfield S 1998 *Phys. Rev. B* **57** 11521

[7] Barnas J and Fert A 1999 *J. Magn. Magn. Mater.* **192** L391

[8] Martinek J, Barnas J, Michalek G, Bulka B R and Fert A 1999 *J. Magn. Magn. Mater.* **207** L1

[9] Takahashi S and Maekawa S 1998 *Phys. Rev. Lett.* **80** 1758

[10] Iwabuchi S, Tanamoto T and Kitawaki R 1998 *Physica B* **249–251** 276

[11] Wang X H and Brataas A 1999 *Phys. Rev. Lett.* **83** 5138

[12] Ono K, Shimada H, Kobayashi S and Ootuka Y 1996 *J. Phys. Soc. Jpn.* **65** 3449

[13] Mitani S, Takahashi S, Takanashi K, Yakushiji K, Maekawa S and Fujimori H 1998 *Phys. Rev. Lett.* **81** 2799

[14] Brückl H, Reiss G, Vinzelberg H, Bertram M, Mönch I and Schumann J 1998 *Phys. Rev. B* **58** R8893

[15] Guérom S, Deshmukh M M, Myers E B and Ralph D C 1999 *Phys. Rev. Lett.* **83** 4148

[16] Nakajima K, Saito Y, Nakamura S and Inomata K 2000 *IEEE Trans. Magn.* **36** 2806

[17] Gittleman J I, Goldstein Y and Bozowski S 1972 *Phys. Rev. B* **5** 3606

[18] Helman J S and Abeles B 1976 *Phys. Rev. Lett.* **37** 1429

[19] Fujimori H, Mitani S and Ohnuma S 1995 *Mat. Sci. Eng. B* **31** 219

[20] Mitani S, Takanashi K, Yakushiji K and Fujimori H 1998 *J. Appl. Phys.* **83** 6524

[21] Fujimori H, Mitani S and Takanashi K 1999 *Mat. Sci. Eng. A* **267** 187

[22] Yakushiji K, Mitani S, Takanashi K, Ha J-G and Fujimori H 2000 *J. Magn. Magn. Mater.* **212** 75

[23] Abeles B, Ping Sheng, Coutts M D and Arie Y 1975 *Adv. Phys.* **24** 407

[24] Imamura H, Chiba J, Mitani S, Takanashi K, Takahashi S, Maekawa S and Fujimori H 2000 *Phys. Rev. B* **61** 46

[25] Takanashi K, Mitani S, Chiba J and Fujimori H 2000 *J. Appl. Phys.* **87** 6331

[26] Yakushiji K, Mitani S, Takanashi K, Takahashi S, Maekawa S, Imamura H and Fujimori H 2001 *Appl. Phys. Lett.* **78** 515

[27] Yakushiji K, Mitani S, Takanashi K and Fujimori H unpublished research

[28] Barner J B and Ruggiero S T 1987 *Phys. Rev. Lett.* **59** 807

[29] Fujii M, Kita T, Hayashi S and Yamamoto K 1997 *J. Phys.: Condens. Matter* **9** 8669

[30] Inoue Y, Tanaka A, Fujii M, Hayashi S and Yamamoto K 1999 *J. Appl. Phys.* **86** 3199

Chapter 16

Ni and Fe nanocrystals

Eiji Kita

Introduction

Nanocrystalline materials have been investigated from fundamental and application viewpoints because of their advantages as novel materials [1]. The size of isolated particles and the grain size in a bulk material often determine the key properties for magnetic material applications. High magnetic permeability is one of the most important features of nanoferromagnets. Nanoferromagnetic materials with soft magnetic properties have been developed by thermal treatment starting from the amorphous state [2, 3], and the magnetic properties have been explained by the random anisotropy model [4]. The random anisotropy model assumes a single element system, whereas thermally crystallized nanostructured materials are composed of at least two phases, crystalline and amorphous. Detailed discussion of the random anisotropy model in a complex system has been reported by Suzuki and co-workers [5].

Ni and Fe nanocrystals prepared by gas condensation and compaction do not contain other phases except for a small amount of entrained oxygen, and their magnetic properties should fit the random anisotropy model [6]. This chapter describes investigations of Ni and Fe nanocrystals manufactured by gas evaporation and deposition [7–9]. The magnetic properties of the 3d ferromagnetic nanocrystals are reported and the random anisotropy model used to explain magnetic anisotropy in the nanocrystals [14].

Chapters 1 and 2 describe the fundamental thermodynamics and structure of nanocrystalline alloys, and chapters 3, 4 and 6–9 describe the manufacture and structure of a number of different nanocrystalline metallic materials.

Random anisotropy model in nanostructured magnets

Soft magnetic characteristics in nanocrystalline materials have been generally interpreted by the random anisotropy model. This model was first introduced

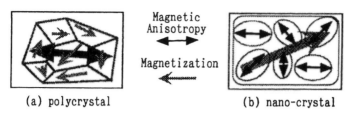

(a) polycrystal   (b) nano-crystal

Figure 16.1. Illustrations of polycrystalline and nanocrystalline magnets. Relationship between sizes of magnetic and crystalline domains is indicated.

to describe the magnetic properties of amorphous ferromagnets composed of rare-earth elements and $3d$ ferromagnetic elements, such as Dy-Fe, where the rare earth metal has strong crystalline anisotropy [10]. This concept was generalized by Chudnovsky *et al* [11] for the case where the coherence lengths of the magnetic coupling and anisotropy are competing. In 1990, Herzer [4] demonstrated that the random anisotropy model can be applied to explain nanocrystalline soft magnetic materials. Figure 16.1 shows schematic illustrations of nanocrystalline and polycrystalline magnets. Figure 16.1(a) shows a polycrystalline substance with an exchange magnetic interaction coherence length larger than the crystal grain size. In this case, the magnetic domains, in which the direction of magnetization is constant, are smaller than the crystalline domains. The magnetization in each domain is subject to an anisotropy energy which is the same as in a single crystal. In contrast, nanocrystals have shorter magnetic exchange coherence length than the crystalline grain size. The anisotropy energy of each magnetic domain is no longer the same as in a single crystal, but is the average contribution from the crystalline domains.

When the magnetic domains include N crystalline domains, the anisotropy is reduced by approximately $1/\sqrt{N}$. The coercive force is then expressed by the relation

$$H_C = P_C \frac{K_r^4 D^6}{M_S A^3} \tag{16.1}$$

where K_r, M_S, A and P_C are the local anisotropy constant, saturation magnetization, exchange stiffness constant and a constant respectively. The coercive force in each nanocrystal region is almost proportional to D^6 and this relationship agrees well with the random anisotropy model.

The dramatic change in H_C can now be interpreted. When the crystal grain size is smaller than the magnetic domain size, rotation dominates the magnetization reversal process and the coercive force varies as D^6 as given by equation (16.1). However, when the crystal grain size is greater than the magnetic domain size, domain wall motion dominates the magnetization reversal process and the coercive force is proportional to D^{-1}.

Preparation of nanocrystalline materials

A variety of physical and chemical fabrication techniques for nanocrystalline materials have been reported. Partial crystallization from an amorphous structure by thermal annealing has been utilized to manufacture nanocrystalline soft magnetic materials. Application of ultrafine particles generated by gas evaporation and condensation have also been used. Bulk material can also be manufactured by gas evaporation and condensation, followed by compaction [1].

Figure 16.2 shows a schematic illustration of the gas evaporation and condensation system. Ultrafine particles are first prepared by evaporating the elementary metals in an inert gas atmosphere. The particles are carried at high speed by helium gas flow under a pressure difference between the evaporation and deposition chambers, to pile up on Si wafer or polyamide film substrates and form a nanocrystalline material. A helium circulation system is installed to provide a high purity helium gas supply at a high rate. Helium gas flow at $40\,l/s$ and $5\,kg/cm^2$ is achieved using two all-metal bellows pumps (MBP in figure 16.1). The helium is then returned to the circulation system and purified by two SAES getter pumps. The oxygen concentration is monitored with a ZrO_2 monitor. The oxygen content is decreased to 10^{-7} torr after 1 h from starting the circulation system.

The pressure difference P_e between the evaporation and deposition chambers and the temperature of the hearth T_h are major factors controlling

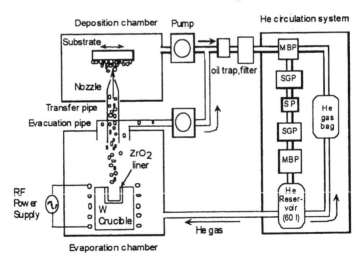

Figure 16.2. Schematic illustration of GDM system for the fabrication of $3d$ transition ferromagnetic nanocrystals. MBP, mechanical bellows pumps; SGP, SAES getter pump; SP, sorption pump.

the material manufacture. Because the evaporation temperature must be higher than the melting point, alloying may take place between the evaporating metal and the evaporation crucible. Ceramic liners are used to prevent such alloying. A ZrO_2 liner is used with a C crucible for Ni, and ZrO_2 liners are also used with W crucibles for Fe and Co.

Ni nanocrystalline samples are suitable in order to study the relation between magnetic properties and grain size. The helium pressure is kept constant in the evaporation chamber at 0.1 MPa (760 torr), and ultrafine particles can be prepared by induction heating Ni metal in a carbon crucible at 1400 °C in a similar manner to inert gas evaporation. After deposition, samples are thermally annealed up to 500 °C in an argon atmosphere to control the grain size.

Nanocrystalline structure

Figure 16.3 shows the change in the x-ray diffraction patterns from Ni nanocrystals annealed at various temperatures. The peaks are identified as fcc, with no diffraction peaks from the oxides. The peak intensities are slightly different from conventional powder patterns, suggesting a $\langle 111 \rangle$ preferred orientation normal to the substrate. A similar trend has been reported in Au nanocrystals, which show strong $\langle 111 \rangle$ orientation [12]. Particle sizes obtained from the width of different diffraction peaks are slightly variable, and averaged values are used as the particle sizes. The grain size of as-deposited nanocrystalline material is about 10 nm and increases to 15 nm with thermal annealing up to 350 °C after film growth.

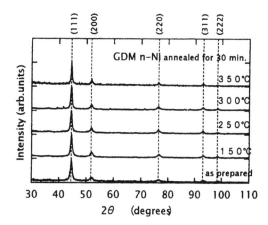

Figure 16.3. X-ray diffraction patterns of NI nanocrystals prepared with GDM. As-prepared samples were thermally annealed up to 350 °C for 30 min.

Nanocrystalline magnetic properties

Magnetic properties are studied mainly by measuring magnetization curves with a vibrating sample magnetometer and a SQUID magnetometer. Thermal annealing with an infrared lamp increases the particle sizes. Inductively coupled plasma photo-emission spectroscopy is used to determine Ni contents and Auger electron spectroscopy is used to investigate oxygen depth profiles.

The magnetization is well saturated at low magnetic fields of approximately 100 Oe. The saturation magnetization for nanocrystalline Fe, Ni and Co prepared with various pressure differences P_e is shown in figure 16.4(a). The hearth temperature was kept constant at values of 1500 °C for Ni, 1950 °C for Fe and 1900 °C for Co (corresponding to different liners and crucibles). In figure 16.4, the saturation magnetizations are almost the same as bulk values, and the dependence on P_e is small. For In nanocrystalline Co, the magnetization is larger than the bulk value by 5–7%. However, clear data proving magnetic enhancement have not yet been obtained.

Figure 16.4(b) shows corresponding coercivity values for the samples in figure 16.4(a). Low coercivity is obtained for all the nanocrystals, in particular approximately 1 Oe is obtained for nanocrystalline Co with $P_e = 1000$ torr. In general, the coercive force decreases with increasing P_e, but it increases to 15 Oe in nanocrystalline Co at 1500 torr, perhaps because of an imperfect fcc crystalline structure. X-ray diffraction results show mainly fcc peaks, with weak additional (100) and (101) hcp peaks. These peaks are particularly clear for the sample made at $P_e = 1500$ torr. The hcp phase seems to be the origin of higher coercive force. When ultrafine particles of Co are produced by gas evaporation and condensation, the high temperature fcc phase is formed. The extent of oxidation in the gas evaporated and deposited

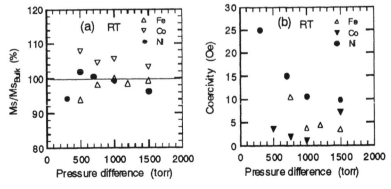

Figure 16.4. (a) Saturation magnetization of *n*-Co, *n*-Fe and *n*-Ni versus the pressure difference P_e. (b) Coercive force of *n*-Co, *n*-Fe and *n*-Ni versus the pressure difference P_e.

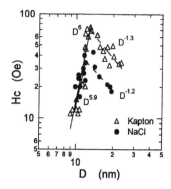

Figure 16.5. Coercive force of Ni nanocrystals prepared with GDM. Solid line in the figure denotes D^6 dependence.

nanocrystals is remarkably small. This is also shown by Mössbauer results from nanocrystalline Fe prepared by gas evaporation and condensation.

Coercive forces in nanocrystalline Ni prepared by gas evaporation and deposition are shown in figure 16.5, and range from 10 to 80 Oe. The data can be fitted to a test function for D^δ giving $\delta = 5.9$, in good agreement with the random anisotropy model [4], as given by equation (1) and as shown in figure 16.5. The exchange stiffness constant is found to be 0.75×10^{-11} 1/m [13]. As the magnitude of P_e ranges roughly from 0.2 to 2 [4], K_r can be estimated to be 1.9–3.5×10^4 J/m^3.

The magnitudes of the local magnetic anisotropy constants are larger than for bulk Ni, 5.7×10^3 J/m^3. There are several possible explanations for the magnitude of the anisotropy. Stresses can induce additional aniso-tropy but, if they are uniaxial, H_C has a D^3 dependence on grain size [5]. Another possibility is that averaging of local anisotropy is weakened by the nature of the interface regions. When the interface regions include oxygen atoms, nonmagnetic impurities and vacancies, these can lead to a decrease in the exchange coherence length and the averaging effect is sup-pressed. Oxide formation at the interface often causes exchange anisotropy. From the preliminary magnetization measurement, H_C increased to 60 Oe at 4.2 K, however the MH loop did not shift after magnetic field cooling from room temperature at 0.5 T. This suggests that the influence of oxygen on the magnetic properties is not large in nanocrystalline material, and supports the low oxide contents incorporated during fabrication. A clear hysteresis loop shift is observed during compaction of the nanocrystals [6]. A final possibility is that a small amount of oxygen atoms between grains may decrease the magnitude of the exchange stiffness A. In this case, the local anisotropy will be of the same order as the bulk, according to equation (1).

In Fe nanocrystals fabricated by gas evaporation and condensation, H_C seems to vary with D^3, which indicates the presence of uniaxial magnetic anisotropy. This effect has been pointed out by Suzuki and co-workers [5]. Further detailed structural analysis and experiments related to magnetic coherence are needed to elucidate these effects clearly.

References

[1] Gleiter H 1989 *Prog. Mater. Sci.* **33** 223

[2] Yoshizawa Y, Ogawa S and Yamauchi K 1988 *J. Appl. Phys.* **64** 6044

[3] Makino A, Inoue A and Mastumoto T 1999 *NanoStruct. Materials* 12 825

[4] Herzer G 1990 *IEEE Trans. Magn.* **MAG-26** 1397

[5] Suzuki K and Cadogan J M 1998 *Phys. Rev.* **858** 2730
 Suzuki K, Herzer G and Cadogan J M 1998 *J. Magn. Magn. Mater.* **177–181** 949

[6] Laffler J P, Meier J P, Doudin B, Ansermet J-P and Wagner W 1998 *Phys. Rev.* **857** 2915

[7] Kashu S, Fuchita E, Manabe T and Hayashi 1984 *Jpn. J. Appl. Phys.* **23** L910

[8] Kita E, Shiozawa K, Sasaki Y, Iwamoto Y and Tasaki A 1996 *IEEE Trans. Magn.* **MAG-32** 4487

[9] Sasaki Y, Hyakkai M, Kita E, Tanimoto H and Tasaki A 1999 *NanoStruct. Mater.* **12** 907

[10] Alben R, Becker J J and Chi M C 1978 *J. Appl. Phys.* 49 1653

[11] Chudnovsky E M, Saslow W M and Soreta R A 1986 *Phys. Rev.* **833** 521

[12] Tanimoto H, Sakai S, Otsuka K, Kita E and Mizubayashi H unpublished research

[13] Kimura R and Nose H 1962 *J. Phys. Soc. Jpn.* Suppl **8–1** 604

[14] Sakai T, Tsukahara N, Tanimoto H, Ota K, Murakami H and Kita E 2001 *Scripta Mater.* **44** 1359

Chapter 17

Nanocrystalline Fe-M-B alloys

Akihiro Makino

Introduction

This chapter reviews results on the development of soft magnetic Fe-M-B (M = Zr, Nb, Hf) nanocrystalline alloys with high saturation magnetic flux density B_s above 1.5 T as well as excellent soft magnetic properties. Industrial applications are discussed in more detail in chapters 18 and 19. A mostly single bcc structure composed of α-Fe grains about 10–20 nm in size surrounded by a small amount of an intergranular amorphous layer is obtained by crystallization of amorphous alloys prepared by melt-spinning. The typical nanocrystalline bcc $Fe_{90}Zr_7B_3$, $Fe_{89}Hf_7B_4$ and $Fe_{84}Nb_7B_9$ ternary alloys subjected to optimum annealing exhibit high saturation magnetic flux density $B_s > 1.5$ T, as well as high effective permeability $\mu_e > 20\,000$ at 1 kHz. Excellent soft magnetic properties of the nanocrystalline Fe-M-B based alloys can be obtained by decreasing the bcc grain size D and magnetostriction λ, and increasing the Curie temperature T_C of the intergranular amorphous phase. The quaternary $Fe_{85.5}Zr_2Nb_4B_{8.5}$ alloy shows high effective permeability $\mu_e = 60\,000$ at 1 kHz, high saturation magnetic flux density $B_s = 1.64$ T and zero-magnetostriction λ simultaneously. However, these alloys have to be produced with a melt-spinning apparatus in a vacuum chamber, which is a great disadvantage in industrial applications. Crystallized $Fe_{84.9}Nb_6B_8P_1Cu_{0.1}$ prepared in air has an as-quenched structure composed of the amorphous phase and nanoscale α-Fe grains, and exhibits higher effective permeability $\mu_e = 41\,000$ at 1 kHz and higher saturation magnetic flux density $B_s = 1.61$ T, as well as a considerably lower core loss of 0.1 W/kg at 50 Hz and 1.4 T, compared with the Fe-based amorphous alloys now used as core materials for transformers.

Background

The development of some kinds of nanocrystalline alloys such as Al-based high-strength alloys [1] and functional materials has been tried recently by crystallizing an amorphous phase. It has already been reported by Masumoto *et al* [2] that an amorphous phase is useful as a precursor to prepare a nanocrystalline structure. In 1981, Koon and Das [3] found that a fine crystallized structure of an Fe-B-Ta-Ca alloy exhibits a rather high coercivity H_C. Subsequently, the crystallized state of an amorphous Fe-Nd-B phase has been found to show good hard magnetic properties in many studies [4].

On the other hand, in soft magnetic materials such as soft ferrites, Nα-Fe and Fe-Si alloys, a reduction in grain size of a crystalline phase was considered undesirable, because of the relationship

$$H_C \propto 1/D \qquad (17.1)$$

where D is the grain size, as proposed by Kersten [5]. However, a breakthrough in soft magnetic materials was reported in 1998 [6] for melt-spun Fe-Si-B amorphous alloys containing small amounts of Cu and Nb. The Fe-Si-B-Nb-Cu amorphous phase transforms to an Fe-Si solid solution with a grain size of about 10 nm during annealing at temperature above the crystallization temperature, and alloys with this nanostructure have excellent soft magnetic properties combined with a rather high saturation magnetic flux density of about $B_s = 1.2$–1.3 T. Similar magnetic properties have been obtained for Fe-M-C (M = V, Nb, Ta) sputtered films consisting of α-Fe and MC particles produced by crystallization of the as-sputtered amorphous phase [7]. Good soft magnetic properties of these nanocrystalline alloys are considered to be caused by a decrease in apparent magnetic anisotropy resulting from a homogeneous microstructure, mainly consisting of nanoscale bcc Fe particles.

In 1990, the present author systematically investigated the relation between crystallizing-induced microstructure and magnetic properties for melt-spun amorphous Fe-M-B (M = Zr, Hf, Nb) alloys with the highest Fe concentration in a number of melt-spun soft magnetic Fe-based amorphous alloys. The Fe-M-B alloys consisted [8, 9] mostly of a single bcc structure with nanoscale grains and exhibited good soft magnetic properties as well as saturation magnetic flux densities values above 1.5 T, much higher than those of the nanocrystalline Fe-Si-B-Nb-Cu alloys [6]. The possibility of synthesizing a nanocrystalline structure in Fe-M-B alloys (M = Zr, Hf or Nb) alloys has also been investigated [10]. The resulting nanocrystalline Fe-Zr-Nb-B-Cu alloys [11, 12] have excellent soft magnetic properties of 1.57 T for saturation magnetic flux density B_s and 16×10^4 for effective permeability μ_e at 1 kHz, at the highest level for any soft magnetic material.

Figure 17.1 shows a classification of typical nanocrystalline soft magnetic alloys produced by crystallization of a melt spun amorphous phase. The alloys can be classified into two groups.

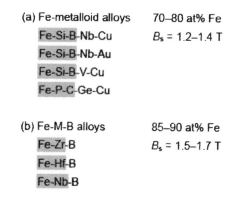

Figure 17.1. A classification of a typical melt-spun nanocrystalline soft magnetic alloys according to the alloys systems.

Group (a) consists of Fe-metalloid-based alloys. It is well known that typical Fe-metalloid amorphous systems such as Fe-Si-B [13] and Fe-P-C [14] show good soft magnetic properties combined with high B_s of about 1.3–1.6 T. The soft magnetic properties of the Fe-based amorphous alloys can be significantly improved by precipitating fine sub-micron size crystalline particles with a long-time annealing treatment at a temperature below that for spontaneous crystallization [15, 16]. The formation of a nanostructure in the Fe-Si-B-Nb-Cu system has been explained [6] by a combination of an accelerated nucleation rate of the bcc phase resulting from the immiscibility of Cu in Fe, and a reduced crystal growth rate due to the small diffusivity of Nb in Fe. However, nanocrystalline Fe-metalloid systems with excellent soft magnetic properties such as Fe-Si-B-Nb-Cu and Fe-P-C-Ge-Cu [17] simultaneously exhibit B_s values lower than Fe-metalloid amorphous alloys mainly because of the decrease in Fe concentration caused by addition of Nb and Cu or Ge and Cu.

Group (b) consists of Fe-M-B based alloys. The formation of an amorphous phase in melt-spun Fe-based alloys usually occurs in the range of 70–84 at% Fe for Fe-metalloid systems [18], and 88–91 at% Fe for Fe-Zr [19] and Fe-Hf [20] systems. It has been reported [21, 22] that addition of small amount of B to Fe-Zr and Fe-Hf amorphous alloys results in an extension of the glass formation range. However, Fe-rich Fe-Zr and Fe-Hf amorphous alloys exhibit an Invar effect [23], accompanied by a decrease in Curie temperature T_C and a decrease in magnetization at room temperature. Hence the B_s values of Fe-Zr and Fe-Hf amorphous alloys are much lower than Fe-metalloid amorphous alloys. If a nanostructure consisting of a metastable bcc solution is formed by crystallization of high Fe content Fe-Zr-B or Fe-Hf-B systems, then the solution is expected to exhibit high B_s. Supersaturated Fe-Zr and Fe-Hf solid solutions prepared by sputtering have low magnetostriction, $\lambda = 1 \times 10^{-6}$. Another new nanocrystalline alloy is Fe-rich

Fe-M-B (M = Zr, Hf, Nb), with high B_s above 1.5 T as well as excellent soft magnetic properties [26].

The development of soft magnetic materials with high B_s combined with good soft magnetic properties continues to be a demand in order to manufacture high performance, miniaturized magnetic parts and devices such as transformers, inductors, magnetic recording heads, and others. From the view point of industrial applications, nanocrystalline Fe-M-B alloys are particularly attractive materials because they have the highest saturation magnetization among the nanocrystalline soft magnetic alloys.

Structure and magnetic properties of Fe-M-B alloys

Figure 17.2 shows the compositional dependence of saturation magnetic flux density B_s, effective permeability μ_e at 1 kHz and saturation magnetostriction λ_s for Fe-Zr-B, Fe-Hf-B and Fe-Nb-B ternary systems subjected to almost optimum annealing at 873–943 K, along with the phase field in the melt-spun state. The maximum effective permeability μ_e in the crystallized state for each alloy is obtained in the compositional range where a fully amorphous phase is formed, and which is close to that where α-Fe is formed.

Figure 17.3 shows the changes in (a) saturation magnetic flux density B_s, (b) effective permeability μ_e at 1 kHz, (c) bcc grain size D, (d) Curie temperature T_C of the residual amorphous phase and (e) saturation magnetostriction λ_s as a function of annealing temperature T_a after 3.6 ks, for typical Fe-M-B alloys which have an amorphous structure in the as-quenched state. The annealed structures are also shown. The saturation magnetic flux densities of the Fe-M-B alloys increase rapidly with increasing annealing temperature in the range from 723 to 773 K, as the structure changes from an amorphous phase with low Curie temperature (resulting from the Invar effect), to a ferromagnetic bcc phase and a small amount of a residual amorphous phase with rather high Curie temperature. In the annealing temperature range above 773 K, the change in saturation magnetic flux density becomes small. The effective permeabilities of the alloys increase with annealing temperature, and reach maximum values of 22 000–32 000 at $T_a = 923$ K, where small grain sizes and nearly zero-magnetostriction are achieved. Transmission electron microscope images for the nanocrystalline alloys annealed at 973 K are shown in figure 17.4. Effective permeability decreases rapidly at $T_a = 950$ K when the bcc phase decomposes to α-Fe + compounds. The rapid decrease in effective permeability is probably explained by the formation of the compounds with large particle sizes which have a larger magnetocrystalline anisotropy, and by the coarsening of the bcc grains which results in an increase in the reduced apparent anisotropy due to the nanoscale grain size.

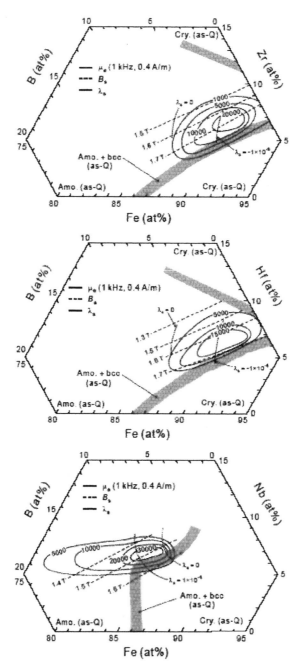

Figure 17.2. Compositional dependence of saturation magnetic flux density (B_s), effective permeability (μ_e) at 1 kH and saturation magnetostriction (λ_s) for Fe-Zr-B, Fe-Hf-B and Fe-Nb-B alloys annealed at 873–943 K, along with the phase field in a melt-spun state.

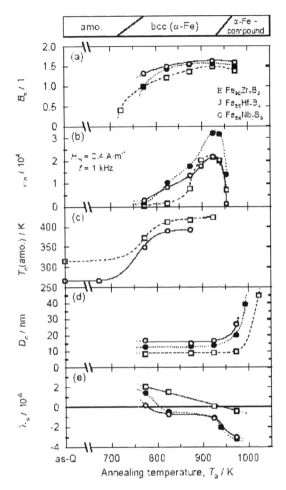

Figure 17.3. Changes in (a) saturation magnetic flux density (B_s), (b) effective permeability (μ_e) at 1 kHz, (c) mean grain size (D), (d) Curie temperature (T_c) of an amorphous phase and (e) saturation magnetostriction (λ_s) as a function of annealing temperature (T_a) for 3.6 ks for amorphous $Fe_{90}Zr_7B_3$, $Fe_{89}Hf_7B_4$ and $Fe_{84}Nb_7B_9$ alloys. The data of annealed structure are also shown.

Structure and magnetic properties of Fe-Zr-Nb-B alloys

The nanocrystalline ternary Fe-M-B alloys exhibit small but non-zero saturation magnetostriction λ_s [27]. It is expected that the soft magnetic properties of the Fe-M-B alloys can be improved further by achieving zero-saturation magnetostriction λ_s. Zero-magnetostrictive Fe-Zr-Nb-B alloys have been studied [27] with a mixed composition of Fe-Zr-B with negative magnetostriction and Fe-Nb-B with positive magnetostriction.

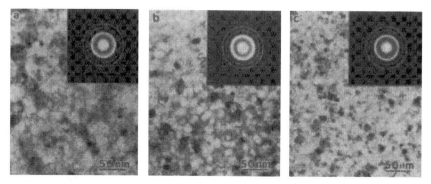

Figure 17.4. TEM images for the crystallized (a) $Fe_{90}Zr_7B_3$, (b)$Fe_{89}Hf_7B_4$ and (c) $Fe_{84}Nb_7B_9$ alloys annealed at 923 K for 3.6 ks.

Figure 17.5 shows a pseudo-ternary diagram of μ_e (solid lines), B_s (broken lines), and λ_s (dotted lines) for Fe-(Zr,Nb)-B alloys crystallized under optimum conditions, with the Zr+Nb content is kept constant at 6 at%. A small grain size of 10–11 nm is obtained in the compositional range 0–3 at% Zr and 6–9 at% B. The permeability reaches a maximum value of 60 000 for the $Fe_{85.5}Zr_2Nb_4B_{8.5}$ alloy, which shows zero-magneto-striction λ_s.

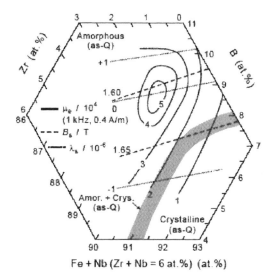

Figure 17.5. Pseudo-ternary diagram of permeability (μ_e), saturation magnetic flux density (B_s) and magnetostriction (λ_s) for nanocrystalline Fe-(Zr,Nb)$_6$B alloys after annealing at the optimum conditions. The data of the phase field in an as-quenched state are also shown.

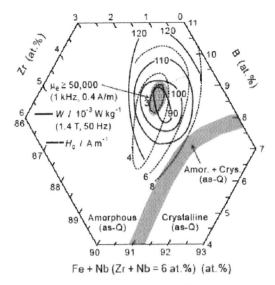

Figure 17.6. Pseudo-ternary diagram of core loss (W) and coercivity (H_c) for nano-crystalline Fe-$(Zr,Nb)_6$B alloys after annealing at the optimum conditions. The grey area indicates the compositional range obtaining high permeability more than 50 000. The data of the phase field in an as-quenched state are also shown.

The core loss W at 1.4 T and 50 Hz of the Fe-(Zr,Nb)-B alloys have been studied to examine their application potential as a core material for pole transformers. Figure 17.6 shows the pseudo-ternary diagram of core loss W (solid lines) and coercivity H_C (broken lines) for Fe-$(Zr,Nb)_6$-B alloys crystallized under optimum conditions. The grey region indicates where high effective permeability μ_e values of above 50 000 have been obtained. An extremely low core loss of less than 0.09 W/kg is obtained in the compositional range 1.5–2.2 at% Zr and 8–9 at% B. The compositional range for minimum core loss W extends to lower boron content compared with the region for best permeability μ_e and coercivity H_C. This is because of an increase in saturation magnetic flux density B_s with decreasing boron content. Since core loss increases rapidly near magnetic saturation (figure 17.7), higher saturation magnetic flux density is favourable for obtaining low core loss. Figure 17.7 shows the core loss at 50 Hz for nanocrystalline $Fe_{85.5}Zr_2Nb_4B_{8.5}$ as a function of maximum induction B_m. The data for nanocrystalline $Fe_{90}Zr_7B_3$, $Fe_{84}Nb_7B_9$ and a commercial amorphous $Fe_{78}Si_9B_{13}$ alloy are also shown. The $Fe_{85.5}Zr_2Nb_4B_{8.5}$ alloy exhibits a very low core loss of 0.09 W/kg at 1.4 T and 50 Hz, which is half to two-thirds that of Fe-M-B ternary alloys and is much lower than the amorphous $Fe_{78}Si_9B_{13}$ alloy. The structure and magnetic properties of the typical ternary and quaternary alloys are summarized in table 17.1.

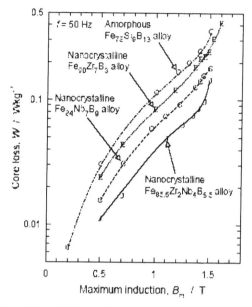

Figure 17.7. Core loss (*W*) at 50 Hz as a function of maximum induction (*B*$_m$) for nano-crystalline Fe$_{90}$Zr$_7$B$_3$, Fe$_{84}$Nb$_7$B$_3$, Fe$_{85.5}$Zr$_2$Nb$_4$B$_{8.5}$ alloys and amorphous Fe$_{78}$Si$_9$B$_{13}$ alloy.

Table 17.1. Mean grain size and magnetic properties of typical ternary and quaternary nanocrystalline Fe-M-B alloys.

Alloy (produced in vacuum)	Grain size *D* (nm)	Magnetic flux density *B*$_s$ (T)	Permeability* (μ)	Coercivity *H*$_C$ (A/m)	Saturation magneto-striction λ_s (10^{-6})	Core loss[†] *W* (W/kg)
Fe$_{90}$Zr$_7$B$_3$	13	1.67	30 000	5.8	−1.1	0.21
Fe$_{84}$Nb$_7$B$_9$	9	1.52	36 000	5.6	0.6	0.15
Fe$_{85.5}$Zr$_2$Nb$_4$B$_{8.5}$	11	1.64	60 000	3.0	−0.1	0.09
Fe$_{84.9}$Nb$_6$B$_8$P$_1$Cu$_{0.1}$	10	1.61	41 000	4.7	—	0.11

* At 1 kHz, 0.4 A/m
[†] At 50 Hz, 1.4 T

Explanation of good magnetic properties

Summarizing the experimental results [10, 27–29], the detailed microstructure, which exhibits good soft magnetic properties, is schematically shown in figure 17.8. The structure is composed mainly of α-Fe grains with a residual intergranular amorphous layer. The nanoscale α-Fe grains contain low transition metal and boron contents, resulting in low magnetostriction.

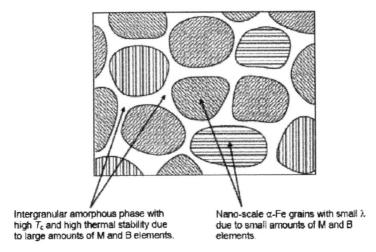

Intergranular amorphous phase with high T_c and high thermal stability due to large amounts of M and B elements.

Nano-scale α-Fe grains with small λ due to small amounts of M and B elements.

Figure 17.8. Schematic diagram of the nanostructure for Fe-M-B alloys.

The intergranular amorphous phase has a rather high Curie temperature resulting from high transition metal and boron contents, causing strong magnetic exchange coupling of the bcc grains. Consequently, good soft magnetic properties should be obtained.

The high saturation magnetic flux density of the alloys originates in their high iron concentrations, higher than those of Fe-metalloid nanocrystalline and amorphous alloys. The dominant factors for excellent soft magnetic properties in the alloys are:

1. a decrease in apparent magnetic anisotropy, resulting from the combined effects of a nanoscale bcc grain size and strong magnetic coupling between the bcc grains through the intergranular ferromagnetic amorphous phase,
2. a small saturation magnetostriction, resulting from a non-equilibrium α-Fe bcc phase with superaturated levels of solute elements in solid solution,
3. a homogeneous structure without coarse particles.

Structure and magnetic properties of Fe-Nb-B-P-Cu alloys produced in air

Typical compositions of Fe-M-B alloys are ternary $Fe_{90}Zr_7B_3$, $Fe_{89}Hf_7B_4$ and $Fe_{84}Nb_7B_9$ and quaternary $Fe_{85.5}Zr_2Nb_4B_{8.5}$ [10, 11, 27–32]. The alloys are produced by melt-spinning in a vacuum or Ar atmosphere because the transition metals have high oxidation activity. The vacuum chamber surrounding the melt-spinning apparatus is a great disadvantage for industrial applications. Production of $Fe_{84}Nb_7B_9$ is easier than $Fe_{90}Zr_7B_3$ and $Fe_{89}Hf_7B_4$ alloys. However, $Fe_{84}Nb_7B_9$ exhibits a saturation

Table 17.2. Melt spinning in air for Fe-Nb-B alloys.

Nb content (at%)	Composition	Melt spinning in air	Structure as-quenched
7	$Fe_{85}Nb_7B_8$	Difficult	—
	$Fe_{84}Nb_7B_9$	Difficult	—
6.5	$Fe_{85}Nb_{6.5}B_{8.5}$	Easy	Amorphous + α-Fe
	$Fe_{84.5}Nb_{6.5}B_9$	Easy	Amorphous + α-Fe
	$Fe_{84}Nb_{6.5}B_{9.5}$	Easy	Amorphous
	$Fe_{83.5}Nb_{6.5}B_{10}$	Easy	Amorphous
	$Fe_{82.5}Nb_{6.5}B_{11}$	Easy	Amorphous
6	$Fe_{83}Nb_6B_{11}$	Easy	Amorphous
	$Fe_{82}Nb_6B_{12}$	Easy	Amorphous
5.5	$Fe_{85}Nb_{5.5}B_{9.5}$	Easy	Amorphous + α-Fe

magnetic flux density of about 1.5 T, the lowest of the typical Fe-M-B alloys. The development of a new nanocrystalline soft magnetic alloy with higher saturation magnetic flux density, above 1.5 T, and with good productivity is, therefore, highly desirable, in order to extend the field of application.

Table 17.2 summarizes the results of melt spinning Fe-Nb-B alloys with various Nb contents in air. $Fe_{84}Nb_7B_9$ samples produced in air are extremely

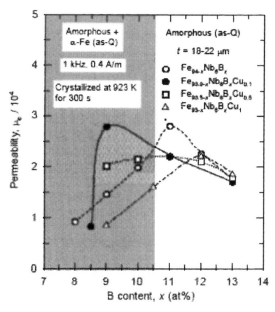

Figure 17.9. Compositional dependence of μ_e in crystallized state of Fe-Nb-B(-Cu) alloys with 6 at% as a function of B content. The as-quenched structure evaluated from the XRD profiles for the alloys is also shown.

oxidized and show brittleness. Ductile ribbons with metallic lustre are obtained for alloys with 6.5 at% Nb or less.

The effect of Cu additions on the as-quenched structure and soft magnetic properties has been studied. Figure 17.9 shows the as-quenched structure evaluated from x-ray diffractometer profiles and the compositional dependence of effective permeability in the crystallized state for Fe-Nb-B-Cu alloys with 6 at% Nb as a function of boron content. For all the alloys, a boron content above 10–11 at% is necessary to obtain a single amorphous structure in the as-quenched state. When the boron content is less than 10–11 at%, a mixed structure of α-Fe and amorphous phases is formed in the as-quenched state. The copper addition scarcely changes the glass-forming ability of the alloys. Figure 17.10 shows x-ray diffractometer profiles taken from the free surface in the as-quenched state of $Fe_{85-x}Nb_6B_9Cu_x$ ($x = 0$, 0.1, 0.5, 1) alloys. A diffraction peak corresponding to α-Fe (110) together with a halo from an amorphous phase are observed in the profiles around $2\theta \approx 52°$ for all the alloys. The values for the mean grain size D of the α-Fe phase roughly estimated only from the α-Fe diffraction peak are about 45, 25 and 50 nm for the Cu-free, 0.5 at% Cu and 1 at% Cu alloy respectively. The grain size of the 0.1 at% Cu alloy is extremely small, but cannot be estimated because the diffraction peak is too broad and unclear.

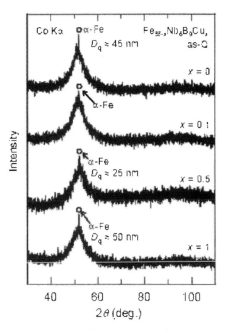

Figure 17.10. XRD profiles taken from free surface in as-quenched state of $Fe_{85-x}Nb_6B_9Cu_x$ alloys ($x = 0, 0.1, 0.5, 1$). Mean grain size (D_q) of α-Fe phase estimated by broadening the diffraction peak is also shown.

This indicates that adding 0.1 at% Cu to $Fe_{85}Nb_6B_9$ suppresses coarsening of the α-Fe grains in the as-quenched state around the glass-formation limit.

As shown in figure 17.9, the maximum effective permeability μ_e for the crystallized alloys without Cu and with 1 at% Cu is obtained at compositions where a single amorphous structure is formed in the as-quenched state. This agrees with previous reports [28, 30] that the precursor to Fe-based nano-crystalline alloys with good soft magnetic properties should be a single amorphous phase. On the other hand, with addition of 0.1 at% Cu, the maximum effective permeability of 28 000 in the crystallized state is obtained at 9 at% B, where an amorphous phase and extremely small α-Fe grains are formed in the as-quenched state. In other words, addition of 0.1 at% Cu changes the as-quenched structure and expands the compositional range where good soft magnetic properties are obtained in the crystallized state at high Fe content.

Phosphorus additions

The addition of P to Fe-Nb-B(-Cu) alloys with 6 at% Nb has been studied to improve the soft magnetic properties of the alloys. Figure 17.11 shows the as-quenched structure and effective permeability μ_e in the crystallized state of

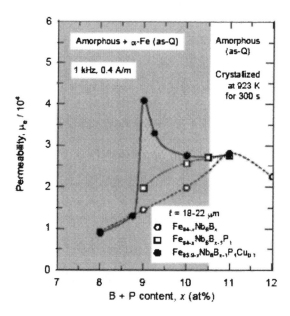

Figure 17.11. Compositional dependence of μ_e in crystallized state of Fe-Nb-B(-P-Cu) alloys with 6 at% Nb as a function of B+P content. The as-quenched structure evaluated from the XRD profiles for the alloys is also shown here.

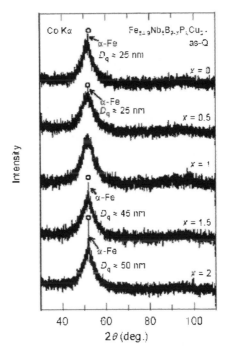

Figure 17.12. Change in XRD profile taken from free surface in as-quenched state of $Fe_{84.9}Nb_6B_{9-x}P_xCu_{0.1}$ alloys as a function of P content. Values of mean grain size (D) of α-Fe phase estimated by broadening the diffraction peak are also shown.

Fe-Nb-B(-P-Cu) alloys as a function of (B+P) content. The effective permeability is improved by replacing 1 at% B by 1 at% P at around 9–10 at% (B+P). The maximum value of permeability is 41 000 obtained in a crystallized $Fe_{84.9}Nb_6B_8P_1Cu_{0.1}$ alloy. This alloy also exhibits a high saturation magnetic flux density B_s of 1.61 T, considerably higher than the $Fe_{84}Nb_7B_9$ alloy.

Figures 17.12 and 17.13 show the changes in x-ray diffraction profile from the free surface in the as-quenched state for $Fe_{84.9}Nb_6B_{9-x}P_xCu_{0.1}$ and $Fe_{85-x}Nb_6B_8P_1Cu_x$ alloys as a function of x, respectively. The diffraction peak corresponding to α-Fe (110) is observed in the profiles for all the alloys except for $Fe_{84.9}Nb_6B_8P_1Cu_{0.1}$. The mean grain size D of the α-Fe phase is estimated to be about 20–60 nm. The as-quenched structure of $Fe_{84.9}Nb_6B_8P_1Cu_{0.1}$ alloy evaluated from the x-ray diffraction profile seems to be a single amorphous phase. However, P has lower glass-forming ability than B in Fe-Nb-P and Fe-Nb-B systems [33] and it is well known that Cu does not improve glass-forming ability in Fe-based alloys [18]. Therefore, it is reasonable to consider that the as-quenched structure of the $Fe_{84.9}Nb_6B_8P_1Cu_{0.1}$ alloy is composed of an amorphous phase and α-Fe

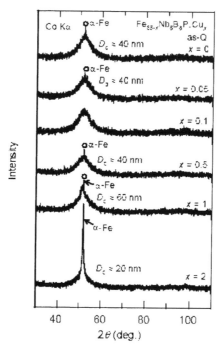

Figure 17.13. Change in XRD profile taken from free surface in as-quenched state of $Fe_{85-x}Nb_6B_8P_1Cu_x$ alloys as a function of Cu content. Values of mean grain size (D) of α-Fe phase estimated by broadening the diffraction peak are also shown.

grains with an extremely small size owing to the simultaneous addition of P and Cu.

Figure 17.14 shows the compositional dependence of the mean grain size in the as-quenched state D_q and the crystallized state D_c, and the effective permeability μ_e in the crystallized state for $Fe_{84.9}Nb_6B_{9-x}P_xCu_{0.1}$ alloys as a function of phosphorus content. The as-quenched structure of the alloys is also shown. The dependence of permeability on phosphorus content agrees qualitatively with the prediction of the random anisotropy model [34], i.e. permeability increases with decreasing grain size. In other words, high permeability results from a small grain size.

Figure 17.15 shows the change of as-quenched and crystallized grain size D_q and D_c and effective permeability μ_e in the crystallized state for $Fe_{85-x}Nb_6B_8P_1Cu_{0.1}$ alloys as a function of copper content. The grain size exhibits a minimum at 1 at% Cu. However, the maximum value of permeability is obtained at 0.1 at% Cu. The 1 at% Cu alloy is composed in the as-quenched state of an amorphous phase and coarse α-Fe grains, about 60 nm in size. The nanocrystallized structure is then composed of grains with an average diameter of 8.8 nm as well as the coarse grains formed in

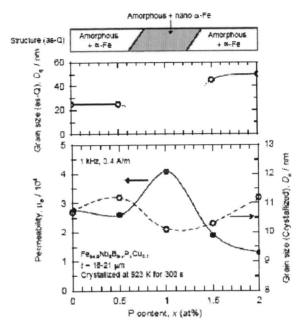

Figure 17.14. Mean grain size in as-quenched state (D_q and crystallized state (D_c, and μ_e in crystallized state of $Fe_{84.9}Nb_6B_{9-x}P_xCu_{0.1}$ alloys as a function of P content. As-quenched structure of the alloys is also shown.

the as-quenched state. On the other hand, the 0.1 at% Cu alloy with the high permeability in the crystallized state has a nanocrystalline structure without coarse grains. It can be concluded that substitution of 1 at% P for B and of 0.1 at% Cu for Fe in $Fe_{85}Nb_6B_9$ is effective in obtaining a nanocrystalline alloy with, simultaneously, good soft magnetic properties, high saturation magnetic flux density B_s and good productivity.

Table 17.1 summarizes the mean grain size and magnetic properties in the crystallized state of the Fe-M-B alloys. The $Fe_{84.9}Nb_6B_8P_1Cu_{0.1}$ alloy exhibits a high effective permeability of 41 000 and a higher saturation magnetic flux density B_s of 1.61 T compared with the $Fe_{84}Nb_7B_9$ alloy, and a lower core loss of 0.11 W/kg at 50 Hz and 1.4 T compared with the $Fe_{84}Nb_7B_9$ and $Fe_{90}Zr_7B_3$ alloys. The $Fe_{84.9}Nb_6B_8P_1Cu_{0.1}$ alloy is more suitable for industrial production because it can be easily produced by a simple rapid-quenching method without any evacuation apparatus.

Concluding remarks

In the present paper, we described recent results on the nanocrystalline soft magnetic Fe-M-B (M = Zr, Hf, Nb) based alloys produced by utilizing the

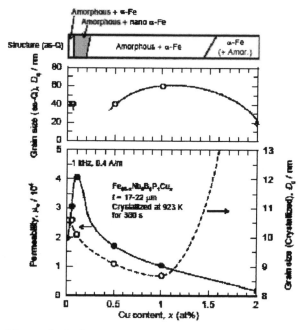

Figure 17.15. Mean grain size in as-quenched state (D_q and crystallized state (D_c, and μ_e in crystallized state of $Fe_{85-x}Nb_6B_8P_1Cu_x$ alloys as a function of Cu content. As-quenched structure of the alloys is also shown.

first stage of crystallization process of the amorphous phase prepared by melt-spinning technique. The crystallized alloys subjected to the optimum annealing have the mostly single bcc structure composed of the nanoscale grain size of about 10 nm surrounded by a small amount of the intergranular amorphous layers with <5 nm in thickness, and exhibit the higher B_s than about 1.5 T, a most attractive property, as well as the excellent soft magnetic properties. The alloys further exhibit the performance of the other important properties, such as thermal stability and stress-sensibility of the good soft magnetic properties, for application. Therefore, the nanocrystalline Fe-M-B based alloys are expected to be used for many kinds of magnetic parts and devices.

References

[1] Kim Y H, Inoue A and Masumoto T 1990 *Mater. Trans. JIM* **31** 747
[2] Masumoto T, Kimura H M, Inoue A and Waseda Y 1976 *Mater. Sci. Eng.* **23** 141
[3] Koon N C and Das B N 1981 *Appl. Phys. Lett.* **39** 840
[4] For example, Croat J J, Herbst J F, Lee R W and Pinkerton F E 1984 *J. Appl. Phys.* **55** 2078
[5] Kersten M 1943 *Z. Phys.* **44** 63

[6] Yoshizawa Y, Oguma S and Yamauchi K 1989 *J. Appl. Phys.* **64** 6044

[7] Hasegawa N and Saito S 1990 *J. Magn. Soc. Jpn.* **14** 313; 1991 *IEEE TJMJ* **6** 120

[8] Suzuki K, Kataoka N, Inoue A, Makino A and Masumoto T 1990 *Mater. Trans. JIM* **31** 743

[9] Suzuki K, Makino A, Inoue A and Masumoto T 1991 *J. Appl. Phys.* **70** 6232

[10] Makino A, Suzuki K, Inoue A, Hirotsu Y and Masumoto T 1994 *J. Magn. Magn. Mater.* **133** 329

[11] Makino A, Inoue A and Masumoto T 1995 *Nanostruct. Mater.* **6** 985

[12] Makino A, Hatanai T, Yoshida S, Hasegawa N, Inoue A and Masumoto T 1996 *Sci. Rep. Res. Inst. Tohoko Univ.* **A42** 121

[13] For example, Masumoto T 1981 *Sci. Rep. Res. Inst. Tohoko Univ.* **A29** 265

[14] Fujimori H, Masumoto T, Obi Y and Kikuchi M 1974 *Jpn. J. Appl. Phys.* **13** 1889

[15] Hasegawam R, Fish G E and Rammanan V R V 1984 *Proc. 4th Int. Conf. on Rapidly Quenched Materials* ed. T Masumoto and K Suzuki (Sendai: Japan Inst. Metals) p 929

[16] Ogata Y, Sawada Y and Miyazaki T 1984 *Proc. 4th Int. Conf. On Rapidly Quenched Materials* ed. T Masumoto and K Suzuki (Sendai: Japan Inst. Metals) p 9

[17] Fujii Y, Fujita H, Seki A and Tomida T J 1991 *Appl. Phys.* **70** 6241

[18] Masumoto T *et al* (eds) 1982 *Materials Science of Amorphous Metals* (Tokyo: Ohmu) p 281

[19] Nose M and Masumoto T 1980 *Sci. Rep. Res. Inst. Tohoko Univ.* **A28** 232

[20] Inoue A, Kobayashi K and Masumoto T 1980 *Proc. Conf. Metallic Glasses, Science and Technology* vol II ed. C Hargitai *et al* (Budapest: Cent. Res. Inst. Phys.) p 217

[21] Ohnuma S, Nose M, Shirakawa K and Masumoto T 1981 *Sci. Rep. Res. Inst. Tohoko Univ.* **A29** 254

[22] Inoue A, Kobayashi K, Nose M and Masumoto T 1981 *J. Phys. C* **8-41** 31

[23] Shirakawa K, Ohnuma S, Nose M and Masumoto T 1980 *IEEE Trans. Mag.* **MAG-16** 910

[24] Katoka N, Hosokawa M, Inoue A and Masumoto T 1989 *Jpn. J. Appl. Phys.* **28** L262

[25] Katoka N, Hosokawa M, Inoue A and Masumoto T 1990 *Mater. Trans. JIM* **31** 429

[26] Makino A, Suzuki K, Inoue A and Masumoto T 1991 *Mater. Trans. JIM* **32** 551

[27] Makino A, Inoue A and Masumoto T 1995 *Mater. Trans. JIM* **36** 924

[28] Makino A, Yamamoto Y, Hirotsu Y, Inoue A and Masumoto T 1994 *Mater. Sci. Eng. A* **179/180** 495

[29] Makino A, Hatanai T, Inoue A and Masumoto T 1997 *Mater. Sci. Eng. A* **226–228** 594

[30] Makino A, Yoshida S, Inoue A and Masumoto T 1994 *IEEE Trans. Mag.* **30** 4848

[31] Makino A, Bitoh T, Kojima A, Inoue A and Masumoto T 2000 *J. Appl. Phys.* **87** 7100

[32] Makino A, Bitoh T, Kojima A, Inoue A and Masumoto T 2000 *J. Magn. Magn. Mater.* **215/216** 288

[33] Bitoh T, Makino A, Ito Y and Tagami M unpublished research

[34] Herzer G 1989 *IEEE Trans. Mag.* **25** 3327

Chapter 18

Advances in nanocrystalline soft magnetic materials

Yoshihito Yoshizawa

Introduction

Nanocrystalline Fe-Cu-Nb-Si-B soft magnetic alloys fabricated by crystallizing amorphous alloys exhibit attractive magnetic properties such as high saturation magnetic induction, high permeability, and low core loss simultaneously [1–3]. Following the author's original paper on nano-crystalline Fe-Cu-Nb-Si-B alloys [1], other nanocrystalline alloys such as Fe-Zr-B(Cu) have been reported with high saturation magnetic induction $B_s = 1.5$–$1.7\,\mathrm{T}$ [4], and an $Fe_{44}C0_{44}Zr_7B_4Cu_1$ alloy has been assessed for high temperature jet engine applications [5, 6]. There is a wide variety of different nanocrystalline soft magnetic alloys, but the Fe-Cu-Nb-Si-B alloy was the first to be commercialized and it has been widely applied to various devices.

The random anisotropy model [7] explains soft magnetic behaviour when the grain size is reduced into the nanoscale range, as described by Herzer [8]. Suzuki *et al* [9, 10] have reported an extended random anisotropy model.

The soft magnetic properties are strongly dependent on grain size, magnetostriction, intrinsic magnetocrystalline anisotropy, exchange coupling, and induced magnetic anisotropy, which change with chemical composition and annealing condition. Structural and magnetic behaviour of nanocrystalline Fe-Cu-Nb-Si-B soft magnetic alloys has been investigated actively. The aims of this chapter are to discuss recent advances in nanocrystalline soft magnetic alloys and to comment on their potential applications. Recent alloying advances are discussed further in chapter 17, and industrial applications are described in more detail in chapter 19.

Amorphous ribbons

Amorphous alloy ribbons of about 18 μm thickness are prepared by a single-roller melt-spinning technique. The as-cast amorphous ribbons are wound into toroidal cores and then annealed under nitrogen with or without a magnetic field. The d.c. B-H loop, relative permeability μ_r at $H = 0.05 \, \text{A m}^{-1}$ and core loss are measured with an automatic hysteresis loop tracer, impedance gain phase analyser and B-H analyser respectively. The crystallization temperature is measured by differential scanning calorimetry. The crystal structure and grain size were examined by x-ray diffraction and the microstructure was observed by transmission electron microscopy.

Advances in nanocrystalline soft magnetic materials

In general, higher saturation magnetic induction and permeability for soft magnetic materials are required to reduce device size. However, the permeability of nanocrystalline Fe-Cu-Nb-Si-B alloys generally decreases with increasing saturation magnetic induction. Hence, it is important from the viewpoint of applications to try to improve the soft magnetic properties of the nanocrystalline Fe-Cu-Nb-Si-B alloys with high saturation magnetic induction. It is well known that:

1. the appropriate copper content is approximately 1 at% for high permeability in the high Si content composition [1, 2],
2. the saturation magnetic induction of nanocrystalline FeCuNbSiB alloys increases with decreasing Si content,
3. the permeability generally decreases in low Si content composition.

Figure 18.1 shows the Cu content dependence of magnetic properties in $Fe_{77}Cu_xNb_{3-x}Si_{11}B_9$ alloys annealed without any magnetic field [11]. The optimum Cu content shifts to $x < 1 \, \text{at}\%$ for high Si content alloys. The maximum value of relative permeability exceeds 150 000 at $x = 0.6 \, \text{at}\%$. Thus, a high saturation flux density B_s of 1.45 T and a high permeability μ_r of over 150 000 are achieved at a Cu content of $x = 0.6 \, \text{at}\%$.

Bright-field transmission electron microscope images of $Fe_{77}Cu_xNb_{3-x}Si_{11}B_9$ alloys ($x = 0$, 0.6, 1.0) are shown in figure 18.2. Randomly oriented grains with diameters of approximately 10 to 15 nm are observed. The average grain size at $x = 0.6 \, \text{at}\%$ is smaller and more homogeneous than that at $x = 1.0 \, \text{at}\%$. Ohnuma *et al* reported that these microstructures are associated with the density of copper clusters during the crystallization of bcc α-Fe primary crystals in the alloy, and the grain size of of bcc α-Fe affects the permeability [12].

The magnetic properties in $Fe_{90.4-a-y}Cu_{0.6}Nb_aSi_yB_9$ alloys have also been investigated to try to improve the soft magnetic properties in

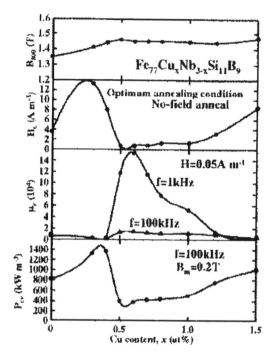

Figure 18.1. Cu content dependence of magnetic properties in $Fe_{77}Cu_xNb_{3-x}Si_{11}B_9$ alloys.

Fe-Cu-Nb-Si-B alloys with high saturation magnetic induction. Figure 18.3 shows the Si content dependence of the magnetic properties in $Fe_{90.4-a-y}Cu_{0.6}Nb_aSi_yB_9$ alloys without any magnetic field. By optimizing the copper and niobium contents, the relative permeability at 1 kHz is about $\mu_r = 105$ and the magnetic induction is high at about 1.50 T in nanocrystalline $Fe_{78.3}Cu_{0.6}Nb_{2.6}Si_{9.5}B_9$ [13].

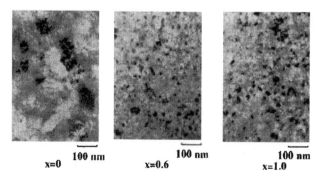

Figure 18.2. Bright-field transmission electron microscopy images of $Fe_{77}Cu_xNb_{3-x}Si_{11}B_9$ alloys ($x = 0$, 0.6, 1.0) annealed at 823 K for 3.6 ks.

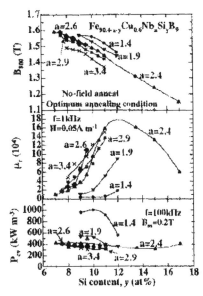

Figure 18.3. Si content dependence of magnetic properties in $Fe_{90.4-a-y}Cu_{0.6}Nb_aSi_yB_9$ alloys.

The magnetic anisotropy induced by a transverse field annealing (HT) improves magnetic properties such as core loss, and quality factor Q in the high frequency range.

A high induced magnetic anisotropy K_u is expected to improve the magnetic properties in the high frequency range. Co addition to $Fe_{78.8}Cu_{0.6}Nb_{2.6}Si_9B_9$ alloy has been investigated to try to increase K_u and improve the magnetic properties in the high frequency range. Figure 18.4

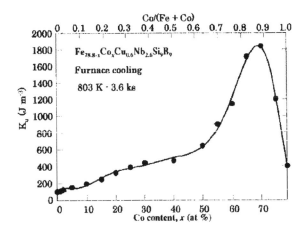

Figure 18.4. K_u as a function of Co content x for $Fe_{78.8-x}Co_xCu_{0.6}Nb_{2.6}Si_9B_9$ alloys.

Table 18.1. Magnetic properties of nanocrystalline Fe-Cu-Nb-Si-B and Fe-Co-Cu-Nb-Si-B alloys.

Composition (at%)	B_{800} (T)	$B_t B_{800}^{-1}$ (%)	H_c (A m^{-1})	μ_r	P_{cv} (kW m^{-3})	Q	K_u (J m^{-3})
$Fe_{77}Cu_{0.6}Nb_{2.4}Si_{11}B_9$	1.45	61	0.8	157 000	380	—	—
$Fe_{78.3}Cu_{0.3}Nb_{2.6}Si_{9.5}B_9$	1.50	40	1.0	109 000	370	—	—
$Fe_{78.8}Cu_{0.6}Nb_{2.6}Si_9B_9$ ($H^{\perp}T$)	1.51	3	2.6	7800	280	0.54	96
$Fe_{73.8}Co_5Cu_{0.6}Nb_{2.6}Si_9B_9$ ($H^{\perp}T$)	1.52	1	1.4	5300	250	0.85	151
$Fe_{8.8}Co_{70}Cu_{0.6}Nb_{2.6}Si_9B_9$ ($H^{\perp}T$)	1.08*	1†	30.3	230	240	13.4	1840
$Fe_{73.5}Cu_1Nb_3Si_{13.5}B_9$	1.24	56	1.1	134 500	290	0.38	—
$Fe_{73.5}Cu_1Nb_3Si_{13.5}B_9$ ($H^{\perp}T$)	1.24	11	0.8	35 000	230	0.67	15

B_{800}: magnetic induction at $H = 800\,\mathrm{A\,m^{-1}}$; μ_r: relative permeability at $f = 1\,\mathrm{kHz}$, $H = 0.05\,\mathrm{A\,m^{-1}}$; P_{cv}: core loss per unit volume at $f = 100\,\mathrm{kHz}$, $B_m = 0.2\,\mathrm{T}$; Q: quality factor at $f = 1\,\mathrm{MHz}$; * B_{8000}: magnetic induction at $H = 8000\,\mathrm{A\,m^{-1}}$; † $B_r B_{8000}^{-1}$.

shows K_u as a function of Co content x for $Fe_{78.8-x}Co_xCu_{0.6}Nb_{2.6}Si_9B_9$ alloys annealed at 803 K for 3.6 ks in a magnetic field. The anisotropy K_u increases with Co content, with a maximum of 1840 J/m^3 at a Co content of $x = 70$ at%, about 19 times higher than for a $Fe_{78.8}Cu_{0.6}Nb_{2.6}Si_9B_9$ alloy.

Table 18.1 summarizes the magnetic properties of nanocrystalline Fe-Cu-Nb-Si-B alloys. Low relative permeability of $\mu_r = 210$ at a frequency of $f = 100\,\mathrm{kHz}$, low remanence ratio, low loss, and $Q = 13.4$ at $f = 1\,\mathrm{MHz}$ are all obtained in the $Fe_{8.8}Co_{70}Cu_{0.6}Nb_{2.6}Si_9B_9$ alloy.

Applications of nanocrystalline soft magnetic materials

Figure 18.5 summarizes the features and typical applications of nanocrystalline Fe-Cu-Nb-Si-B alloys. From features such as high saturation flux density B_s, high permeability μ and low core loss, nanocrystalline soft magnetic materials are used for applications such as common mode choke coils for noise filters, pulse transformers, current transformers, choke coils, magnetic switches, and so on.

Figure 18.6 shows the applied magnetic field dependence of shielding for various shield sheets. A shield sheet of nanocrystalline Fe-Cu-Nb-Si-B alloy is composed of nanocrystalline alloy strips sandwiched by two polymer films with adhesive layers. The shielding effect of the nanocrystalline alloy sheet is larger than for other shield sheets. Hence, the nanocrystalline Fe-Cu-Nb-Si-B alloy sheet is promising as thin and light shield sheets.

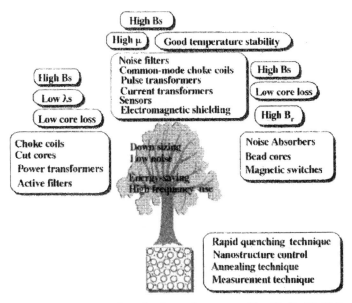

Figure 18.5. Features and typical applications of nanocrystalline Fe-Cu-Nb-Si-B alloys.

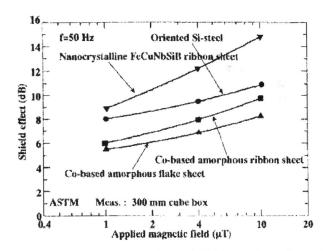

Figure 18.6. Applied magnetic field dependence of shield effect for various shield sheets.

As a new application for nanocrystalline Fe-Cu-Nb-Si-B alloy powder, wave absorber sheet has been developed. Wave absorbers made from nanocrystalline Fe-Cu-Nb-Si-B powder–polymer sheet suppress electromagnetic noise generated by electronic equipment, because of high complex permeability in the high frequency range.

Summary

In summary, the key requirements to develop excellent soft magnetic properties in nanocrystalline materials are to control grain size and phases, by choosing appropriate additives, composition and annealing treatments. Induced anisotropy controlled by magnetic field annealing leads to magnetic properties suitable for applications. By optimizing alloy composition and annealing conditions, high performance nanocrystalline soft magnetic materials have been developed for various applications. Nanocrystalline soft magnetic materials will continue to be used in various applications and will continue to improve their magnetic properties.

References

[1] Yoshizawa Y, Oguma S and Yamauchi Y 1988 *J. Appl. Phys.* **64** 6044
[2] Yoshizawa Y and Yamauchi K 1990 *Mater. Trans. JIM* **31** 307
[3] Yoshizawa Y and Yamauchi K 1991 *Mater. Sci. Eng. A* **133** 176
[4] Suzuki K, Kataoka N, Inoue A, Makino A and Masumoto T 1990 *Mater. Trans. JIM* **31** 743
[5] Willard M A, Laughlin D F, McHenry M E, Thomas D, Sickafus K, Cross J O and Harris V G 1998 *J. Appl. Phys.* **84** 6773
[6] Willard M A, Laughlin D E and McHenry M E 2000 *J. Appl. Phys.* **87** 7091
[7] Alben R, Becker J J and Chi M C 1978 *J. Appl. Phys.* **49** 1653
[8] Herzer G 1989 *IEEE Trans. Magn.* **25** 3327
[9] Suzuki K, Herzer G and Cadogan J M 1998 *J. Magn. Magn. Mater.* **177–181** 949
[10] Suzuki K and Cadogan J M 1998 *Phys. Rev. B* **58** 2730
[11] Yoshizawa Y 1999 *Mater. Sci. Forum* **51** 307
[12] Ohnuma M, Hono K, Linderoth S, Pedersen J S, Yoshizawa Y and Onodera H 2000 *Acta Mater.* **48** 4783
[13] Yoshizawa Y 2001 *Scripta Mater.* **44** 1321

Chapter 19

Applications of nanocrystalline soft magnetic materials

Rainer Hilzinger

Introduction

High performance soft magnetic materials are characterized first by their high permeability μ_i, low coercive force H_c and low losses. A high permeability level requires easy rotation of magnetization, as well as easy domain wall motion, resulting in the demand for low crystalline anisotropy energy K_1 and low magnetostriction λ_s. A low magnetic anisotropy K gives rise to easy magnetization up to saturation and low domain wall energy, i.e. small pinning forces and lowest coercivity H_c. A small magnetostriction λ_s makes the material insensitive to stresses from handling and prevents noise during a.c. magnetization. Applications may also require a high saturation flux density B_s or, at high frequencies, low eddy current losses with the need for high-resistivity. All these different requirements limit the alloy selection and there are only few classes of materials which fulfil both conditions, $\lambda_s = 0$ and $K_1 = 0$.

Table 19.1 gives an overview of the basic material properties governing the magnetic behaviour.

In the past two decades, conventional crystalline materials have been supplemented by amorphous and nanocrystalline alloys. In particular, the newly developed nanocrystalline alloys have a unique combination of properties combining the highest permeability $_i$ with high saturation flux density $B_s \geq 1.2$ [1–4]. Chapters 17 and 18 describe nanocrystalline alloy magnetic properties in detail, whereas this chapter concentrates on industrial applications.

Depending on the relevant application, parameters materials may be selected according to their basic magnetic parameters.

Table 19.1. Basic properties of high permeability magnetic materials.

	Saturation flux density B_s (T)	Curie temperature T_C (°C)	Anisotropy energy K_1 (J/m³)	Magneto-striction λ_s (10^{-6})	Electrical resistivity ρ ($\mu\Omega\,m$)
Permalloy $Ni_{80}Fe_{15}(CuMo)_5$	0.75–0.8	360–400	<100	<1	0.60
Sendust Fe-5.5%Al-9.5%Si	1.2	480	100	<1	0.8
Amorphous Co alloy $Co_{70}Fe_5(Si,B)_{25}$	0.5–1.0	200–400	0	<0.2	1.4
Ferrite $MnO_{0.3}ZnO_{0.2}(Fe_2O_3)_{0.5}$	0.45	120–220	100	1	200 000
Nanocrystalline alloy $Fe_{73}Cu_1Nb_3Si_{16}B_7$	1.3	600	1	<1	1.1

Properties of amorphous and nanocrystalline materials

Nanocrystalline Fe-Cu-Nb-Si-B materials are produced by rapid solidification as thin, initially amorphous ribbons with thicknesses in the range 20–25 μm. The ribbons are wound on toroidal cores and then field-annealed to design the shape of the hysteresis loop according to customer requirements. The inherent thin ribbon gauge and the high electrical resistivity of about 110 μΩ cm reduce eddy current losses even in the higher frequency range. Moreover, for high frequency applications these materials are preferentially used with a flat hysteresis loop. Magnetization thus occurs only via coherent rotation and without domain wall motion. This gives rise to a linear hysteresis loop with almost constant permeability and lowest losses.

Figure 19.1 shows the coercivity H_c as a function of grain size D. For large grain sizes, the coercivity follows a $1/D$ dependence, which reflects the classical rule for conventional polycrystalline materials. Below a grain size of 0.1 μm, however, the coercivity decreases dramatically, following a D^6 law, and for a grain size of 10 nm the coercivity can be as low as for the best permalloy or Co-based amorphous alloy [2, 3]. This range of grain sizes is not directly accessible. The nanocrystalline material is produced initially as an amorphous ribbon. The nanocrystalline state is then achieved by subsequent heat treatment at temperatures of 500–600 °C, i.e. slightly above the crystallization temperature of the amorphous material.

An additional advantage of nanocrystalline Fe-Cu-Ni-Si-B alloys is that the initially high positive saturation magnetostriction λ_s in the amorphous alloy decreases sharply on formation of the nanocrystalline λ_s structure, as shown in figure 19.2. With decreasing the initial permeability μ_i increases to values higher than 100 000. The exact value of magnetostriction λ_s in

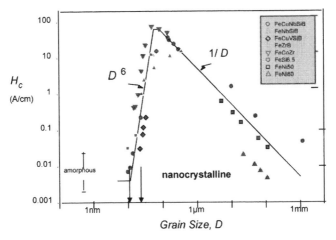

Figure 19.1. Grain size dependence of coercivity.

the nanocrystalline state is mainly determined by the composition, decreasing with Si content and passing zero at a Si content of about 16 at%. This reflects the fact that the major constituent phase of these nanocrystalline materials is Fe-Si, which is well known to exhibit low or vanishing magnetostriction at high Si content. By careful control of alloy composition and annealing treatment, zero magnetostriction nanocrystalline alloys can be produced which have higher saturation flux density than any other high permeability material and which moreover are based on inexpensive raw materials such as Fe and Si.

Figure 19.3 shows the magnetic permeability $\mu(f)$ versus frequency f for various high permeability core materials, such as high permeability ferrite,

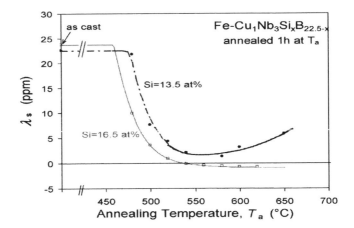

Figure 19.2. Magnetostriction.

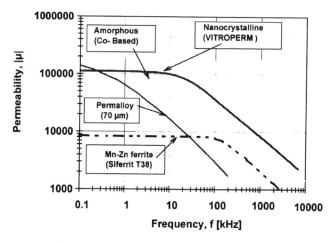

Figure 19.3. Comparison of permeability as a function of frequency.

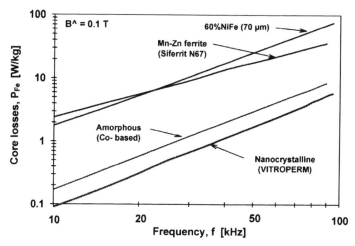

Figure 19.4. Comparison of material losses.

amorphous Co-based alloys and nanocrystalline alloys. Figure 19.4 gives a comparison of core losses at $B = 0.1$ T in the frequency range 10–100 kHz.

Switched mode power supply applications

Depending on the output power, switched mode power supplies operate in the frequency range from about 10 kHz up to several 100 kHz. Various essential magnetic components are used in switched mode power supplies, such as common mode RFI chokes as filters on the input side, power transformers, magamps for independent regulation of output voltages, storage chokes, and spike killers to suppress output voltage noise.

Common mode chokes for RFI filters

Common mode RFI chokes serve to attenuate circuit interference. The attenuation properties are determined by the impedance curve across the interference spectrum. The main material requirements for common mode chokes are as follows:

1. high magnetic permeability μ_i in the range 10 kHz to 10 MHz,
2. high saturation flux density B_s to prevent saturation effects at high noise amplitude or at unbalanced currents,
3. high thermal stability up to 150 °C.

As demonstrated in figure 19.3, the above requirements for common mode chokes are best fulfilled with nanocrystalline cores.

Figure 19.5 gives as examples the 50 Ω attenuation level achieved in the frequency range 10 kHz to 100 MHz. The nanocrystalline alloy shows a much better attenuation level, in particular in the lower frequency range [6]. This material advantage allows a more than 50% reduced build volume and smooth attenuation characteristics compared with a ferrite core. The versatility of these materials/cores enables application specific filter layout.

Magnetic amplifier cores

Invented more than 15 years ago, cores for magnetic amplifiers are now produced mainly of zero magnetostrictive amorphous cobalt based alloys. Magnetic amplifiers have a large market, in particular for power supplies

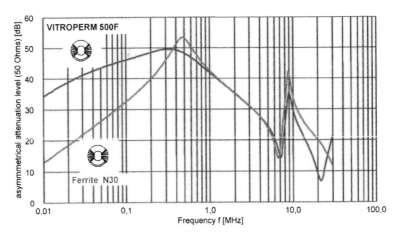

Figure 19.5. Common mode EMC filter chokes. Improved attenuation level.

in personal computers and servers. Nanocrystalline alloys have recently been suggested as suitable materials for these magamps [4].

Kilowatt transformers

Due to new power semiconductors such as IGBT, switched mode power supplies with higher clock rates have been established in the upper power range. In these circuits the power transformer is a key component affecting build volume and efficiency of the switched mode power supply. The main material requirements for kilowatt transformers are as follows:

1. the lowest core losses P_{Fe},
2. high saturation flux density B_s,
3. high temperature stability.

Again the nanocrystalline material turns out to be the preferred choice, exhibiting exceptionally low losses in the relevant frequency range of several tens of kHz (see figure 19.4), constant pulse permeability, better temperature stability and safer operation [5]. User benefits with nanocrystalline materials are:

1. reduced build volumes,
2. reduced weight,
3. very stable temperature behaviour,
4. very high transmittable power of up to 500 kW with a single transformer.

An example at 250 kW is given in figure 19.6. There is a wide spectrum of applications for such In switched mode power supply kilowatt transformers, as shown in figure 19.7.

Open type construction for applications with forced air cooling (v = 2m/s)

Example:
IPT (R)
Induction
Power Transfer

Technical data:
P   = 250 kW,   Output : 1000V/250A   Width   : 300 mm
f   = 15 kHz,   m   : 32,5 kg   Depth   : 150 mm
m/P = 135 g/kW   Height : 175 mm

Figure 19.6. Application example 250 kw power transformer.

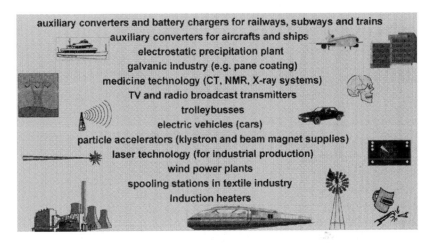

Figure 19.7. Wide spectrum of applications for SMPS-kW-transformers.

Telecommunications applications

In the past decade a new application area has emerged in digital telecommunications, with an increasing demand for interface transformers. ISDN-S_0 signal transformers are used to connect the network termination to consumer terminal equipment such as telephones, telefaxes, personal computers etc. (see figure 19.8). To achieve fidelity of pulse transmission each side of the interface requires two signal transformers incorporating galvanic separation. The transformers must meet very stringent technical requirements, such as compact design and fulfilling the pulse mask even under d.c. premagnetization [4].

Depending on the d.c. premagnetization conditions various amorphous Co alloys or nanocrystalline alloys are used, mainly because of their high magnetic permeability μ_i at frequencies over 10 kHz. Because of their versatile magnetic properties, the amorphous and nanocrystalline Co alloys are again a preferred choice for very high bit rate telecommunication applications such as HDSL, VDSL and ADSL. Broadband transformers and filters for these technologies are already introduced in the market. Figure 19.9 shows as an example an ADSL pots splitter filter.

Cores for ground fault interruptors

Earth leakage circuit breakers for human protection must interrupt electric power for fault currents of 30 mA, both for sinusoidal and pulsating currents. Because of low fault currents without electronic amplification, the

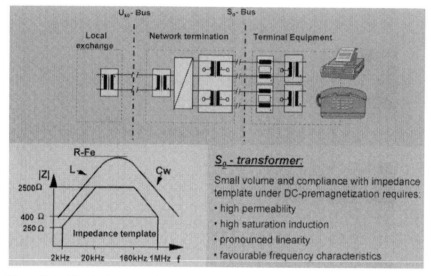

Figure 19.8. ISDN telecommunication-system.

transformer cores must be very sensitive. For pulse currents, a flat hysteresis loop is also essential. Figure 19.10 shows the operating principle and resonance behaviour of a passive pulse current sensitive earth leakage breaker. The nanocrystalline core transfers higher power to the magnetic relay than the best permalloy cores [5, 8]. Nanocrystalline cores of VITRO-PERM 800 F are on the way to substituting for conventional permalloy cores in this application field.

Current transformers for electronic watt hour meter

Electronic watt hour meters are increasingly replacing electromechancal meters in industrial as well as domestic applications. The quality of the meter is determined by the linearity of the current transformer. Nano-crystalline current transformers for electronic watt hour meters offer [7]:

1. negligible amplitude error due to low core losses,
2. small and linear phase error, making compensation easy,
3. very little variation with temperature (see figure 19.11),
4. small build volume.

Conclusions

Zero magnetostrictive nanocrystalline alloys are among the materials with highest permeability and lowest core losses in the frequency range up to

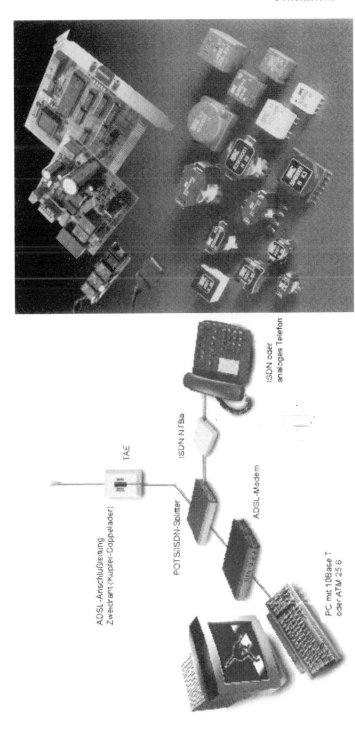

Figure 19.9. Telecommunication. ADSL filters.

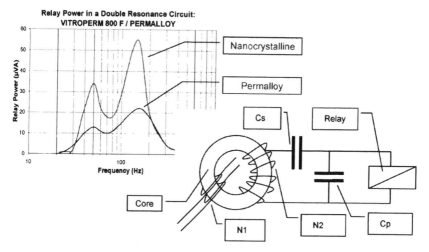

Higher Resonance Power by Lower Magnetic Losses

Figure 19.10. Mains independent residual current devices.

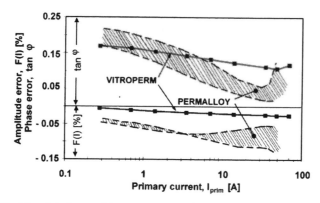

Figure 19.11. Materials for precision current transformers.

1 MHz. The combination of high saturation flux density, $>1.2\,T$, and the highest permeability is unique. By optimizing composition and annealing treatment these alloys can be designed to suit a broad variety of application fields. Core sizes for typical applications range from less than 1 g in telecommunications to many kg for power transformers. Because of these superior properties, nanocrystalline materials are replacing conventional materials to an increasing extent. The use of nanocrystalline flakes consolidated to magnetic cores offers further unique design possibilities, e.g. as moulded chokes for d.c./d.c. power converters [4].

References

[1] Y Yoshisawa, S Oguma and Yamauchi K 1988 *J. Appl. Phys.* **64** 6044

[2] Herzer G 1989 *IEEE Trans. Magn.* **25** 3327; 1990 **26** 1397

[3] Herzer G 1997 in *Handbook of Magnetic Materials* vol 10, ed. K H J Buschos (Elsevier Science)

[4] Petzold J 2004 *Joint European Magnetic Symposia JEMS'01, Grenoble 2001*

[5] Petzold J 1998 *J. Phys. IV France* **8** Pr2-767

[6] Petzold J and Klinger R 2000 *PCIM Europe* **6** 64

[7] J Petzold 1998 *Proc. 37th Int. Conf. Power Conversion PCIM'98* May pp 105–110

[8] European Patents EP 392 204, EP 563 606

Chapter 20

Nd-Fe-B nanocomposite permanent magnets

Satoshi Hirosawa

Introduction

Permanent magnets are essential components of everyday life, used in devices that convert energy. For instance, the majority of battery-driven actuators use permanent magnet direct-current (d.c.) motors. Permanent magnet materials are, therefore, key materials in a modern society where high energy efficiency and low environmental load are priority issues.

The concept of nanocomposite permanent magnets has been a guiding principle in the search for next-generation permanent magnets over the past decade. The first typical nanocomposite permanent magnets based on Nd-Fe-B were reported by researchers at the Philips Research Laboratories [1]. The materials consisted of a mixture of Fe_3B, $Nd_2Fe_{14}B$ and a small amount of α-Fe. The content of Nd in these magnets was about one third of conventional $Nd_2Fe_{14}B$-based magnets and hence a lower material cost was anticipated. Later, 'exchange-spring' behaviour (a semi-reversible demagnetization behaviour originating from nearly reversible rotation of magnetic moments in the magnetically softer components) was noted, and the term 'exchange-spring magnet' was coined by Kneller and Hawig [2]. More recently, Skomski and Coey [3] pointed out the possibility of a textured nanocomposite with giant magnetic energy products exceeding $1\,MJ/m^3$, which is about 2.5 times greater than the record value of $444\,kJ/m^3$ achieved in the best $Nd_2Fe_{14}B$-based [4]. These experimental and theoretical findings have kindled intense research on nanocomposite permanent magnets.

The goal of these researches is twofold: the first is to find and optimize processing routes for the new category of permanent magnets as well as their compositions in order to obtain the desired nanocrystalline micro-structure; the second is to improve their magnetic properties in order to surpass conventional materials. This chapter focuses on processing and properties of nanocomposite permanent magnets based on the Nd-Fe-B system.

Nanocomposite permanent magnet structure

The basic concept of nanocomposite permanent magnets is to combine a large magnetization M_s in a soft metallic magnetic phase and a large magneto-crystalline anisotropy K_1 in a second, hard magnetic phase. Table 20.1 shows the magnetic properties of typical ferromagnetic phases which may be used in a nanocomposite magnets. In an Fe-based system, large magnetization can be obtained by using α-Fe or Fe_3B. The magnetic hardening component can be selected from known hard magnetic phases such as $SmCo_5$, $Sm_2Fe_{17}N_3$ and $Nd_2Fe_{14}B$. However, $Nd_2Fe_{14}B$ is practically the only phase which can coexist in the fully developed crystalline α-Fe or Fe_3B phase of large magnetization. Thus, α-Fe/$Nd_2Fe_{14}B$ and Fe_3B/$Nd_2Fe_{14}B$ nanocomposites have been most intensively studied.

A simple macroscopic mixture is not suitable for combining the large magnetization and magnetic hardening phases because the demagnetization curve of such a mixture would not be useful, as shown in figure 20.1. According to theoretical investigations [4, 5], the grain size of the magnetically softer component in a nanocomposite magnet must be around 10 nm so that the magnetic moments of this component can have a strong intergranular exchange coupling with the hard magnetic component, for instance, $Nd_2Fe_{14}B$. In such a structure, magnetization vectors in the soft magnetic phase are effectively locked to prevent unfavourable rotation of magnetization from occurring. Then the demagnetization curve can become smooth with a single coercivity value.

Table 20.1. Typical ferromagnetic phases for nanocomposite permanent magnets and their magnetic properties at room temperature.

Function in nanocomposite	Phase	Curie temperature T_C (K)	Spontaneous magnetic polarization J_s (T)	Magnetocrystalline anisotropy constant K_1 (MJ/m^3)
Large magnetization	α-Fe	1044	2.15	0.05
	Co	1390	1.81	0.53
	Fe_3B	786	1.6	−0.3
Magnetic hardening	$Nd_2Fe_{14}B$	585	1.61	4.9
	$Sm_2Fe_{17}N_3$	749	1.54	8.6
	$SmCo_5$	1020	1.07	17.2

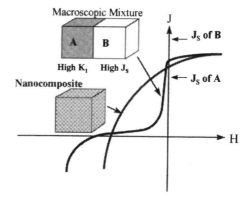

Figure 20.1. Schematic illustration of demagnetization curves of two-phase magnets.

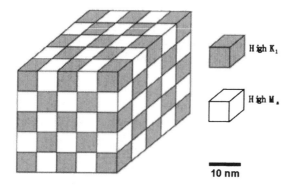

Figure 20.2. Illustration of 'ideal' structure of a nanocomposite permanent magnet.

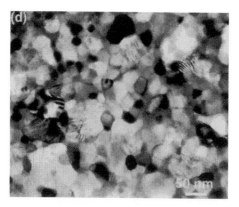

Figure 20.3. Example of microstructure of an isotropic nanocomposite permanent magnet. The material is composed of Fe_3B soft magnetic phase and $Nd_2Fe_{14}B$ hard magnetic phase.

Such a strong magnetic coupling is realized only when there exist atomic contacts among the component phases and the exchange coupling takes place between atomic moments belonging to different phases. Since the coupling energy scales with the area of the interfaces formed between the two magnetic components, a nanocomposite permanent magnet must have a structure such as schematically illustrated in figure 20.2. In the ideal structure, the easy direction of magnetization of all the hard magnetic grains is textured. Figure 20.3 shows an example of the microstructure of a nanocomposite permanent magnet composed of Fe_3B and $Nd_2Fe_{14}B$. In contrast to the ideal structure shown in figure 20.2, the structure shown in figure 20.3 is obviously inhomogeneous and there exists no texture, and the material is isotropic.

Nd-Fe-B nanocomposite permanent magnet composition

Numerous investigations have been performed on nanocomposite permanent magnets based on various combinations of soft and hard magnetic phases. Most of these investigations have dealt with isotropic nanocomposites in which all the grains are randomly oriented. Table 20.2 lists some examples of isotropic nanocomposite magnets and their magnetic properties. Figure 20.4 summarizes B_r and H_{cJ} values to show that these magnets have magnetic properties comparable with nearly-single phase $Nd_2Fe_{14}B$ powder (for instance, MQP-B available from Magnequench International), which is currently commercially produced for resin-bonded magnets. Therefore, there

Table 20.2. Typical magnetic properties of isotropic nanocomposite permanent magnets.

Composition	Ref.	Preparation method	B_r (T)	H_{cJ} (kA/m)	$(BH)_{max}$ (kJ/m^3)
$Nd_4Fe_{80}B_{20}$	[1]	MS-A	1.20	191	93.1
$Nd_{4.5}Fe_{73}Co_3Ga_1B_{18.5}$	[19]	MS-A	1.21	340	128
$Nd_{3.5}Dy_1Fe_{73}Co_3Ga_1B_{18.5}$	[19]	MS-A	1.18	390	136
$Nd_{5.5}Fe_{66}Cr_5Co_5B_{18.5}$	[13]	MS-A	0.86	610	96.6
$Nd_9Fe_{85}B_6$		MS	1.10	485	158
$Nd_7Fe_{89}B_4$		MS	1.28	252	146
$Nd_8Fe_{87.5}B_{4.5}$		MS	1.25	$\sim$500	185.2
$Nd_{3.5}Fe_{91}Nb_2B_{3.5}$		MS	1.45	215	115
$Nd_9Fe_{72.5}Co_{10}Zr_{2.5}B_6$		MS	0.89	$\sim$640	130
$(Nd_{0.95}La_{0.05})_{11}Fe_{66.5}Co_{10}Ti_2B_{10.5}$		MS-A	0.94	1282	146
$Sm_8Zr_3Fe_{85}Co_4N_x$		MS-A-N	0.94	764	118
$Sm_{11.67}Co_{58.33}Fe_{30}$		MA-A	0.97	600	101

MS = melt-spinning, MA = mechanical alloying, A = annealing and N = nitrogenation.

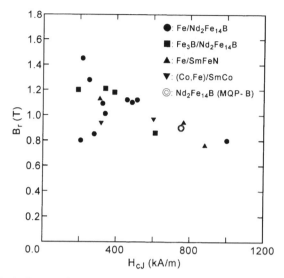

Figure 20.4. Typical magnetic properties of isotropic rare-earth nanocomposite permanent magnets.

is a practical potential for these materials to be utilized in the same application. Shown in this figure are $Fe/Nd_2Fe_{14}B$-type, $Fe_3B/Nd_2Fe_{14}B$-type, Fe/Sn-Fe-N-type and Co-Sm-Co-type.

Nd-Fe-B nanocomposite processing

For the same magnetic performance, what counts is the processing cost of the material. Materials in the Nd-Fe-B system are most promising because their production processes are simple and highly productive in comparison with other systems such as Sm-Fe-N. Typical composition regions of Nd-Fe-B nanocomposite magnets are shown in figure 20.5.

The α-$Fe/Nd_2Fe_{14}B$-type nanocomposites consist mainly of thermodynamically stable phases, α-Fe and $Nd_2Fe_{14}B$, and minor amount of extraneous phases such as an amorphous or iron–boron phase. The $Fe_3B/Nd_2Fe_{14}B$-type nanocomposites consist mainly of metastable Fe_3B. The hard magnetic phase $Nd_2Fe_{14}B$ is also metastable in this concentration range, so that the materials are entirely metastable. The typical composition range of $Fe_3B/Nd_2Fe_{14}B$ nanocomposites is 3–5 at% Nd and 18–19 at% B, the balance being Fe. When the Nd concentration exceeds this range, the formation of metastable $Nd_2Fe_{23}B_3$ prevails. This phase decomposes to form the thermodynamically stable phases Fe and $Nd_{1.1}Fe_4B_4$ in Nd-rich ternary alloys. When this occurs, good hard magnetic properties will not

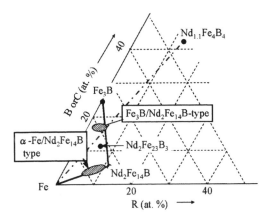

Figure 20.5. Typical compositional regions for α-Fe/Nd$_2$Fe$_{14}$B and Fe$_3$B/Nd$_2$Fe$_{14}$B nano-composite permanent magnets.

be obtained because of the presence of too much soft magnetic and non-magnetic phases.

The most practical method to produce nanostructured metallic materials is rapid solidification. The melt-spinning method has been used in production of isotropic Nd$_2$Fe$_{14}$B-based sub-microcrystalline alloys used for resin-bonded permanent magnets applications. In this method, a melt is ejected through a narrow orifice on to a rotating substrate or a wheel which acts as a chill block. In an industrial melt-spinner or a jet-caster, the melt is stored in a vessel which has an orifice at the bottom and the level of the melt is maintained within a controlled range so that the ejection pressure is maintained practically constant. The rotating wheel needs to be cooled continuously by running water inside in order to enable a continuous operation.

The features of the Nd-Fe-B alloys relevant to the rapid solidification process are as follows. In the Fe-rich region relevant to α-Fe/Nd$_2$Fe$_{14}$B, the liquidus plane corresponds to primary solidification of γ-Fe, which transforms to α-Fe at lower temperatures, below 1192 K. The amorphous formability of alloys in this composition region is poor. On the other hand, the alloy compositions of the Fe$_3$B/Nd$_2$Fe$_{14}$B nanocomposites fall in the region of a deep valley of the liquidus plane. Accordingly, the amorphous formability of these alloys is excellent. This feature is industrially important because it allows a large productivity in the rapid solidification process.

Nanocomposite formation during rapid solidification

Preparation of a nanocomposite permanent magnet by means of crystallization of a rapidly solidified amorphous alloy involves crystallization of multiple

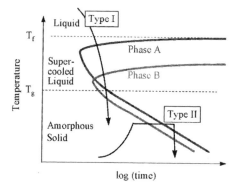

Figure 20.6. Concept of time-temperature-transformation (TTT) curves for crystallization of composite phases (A and B) in rapid solidification of a nanocomposite permanent magnet alloy and the two ways to produce a nanocomposite: Type I is the direct method and Type II is the two-stage method. T_f, liquidus temperature; T_g, glass-transformation temperature.

phases, which may be described by time-temperature transformation (TTT) or continuous cooling transformation (CCT) curves. Figure 20.6 shows schematic TTT curves for a nanocomposite permanent magnet. By rapid solidification, the magnet can be processed in two ways. The first is a direct method in which the two constituent phases crystallize during rapid solidification (type I). The second is a two-step method in which an amorphous or over-quenched precursor alloy is crystallized by heating (type II).

In α-Fe-based systems, the liquidus is at about 1500 K and the peritectic temperature for $Nd_2Fe_{14}B$ formation is about 1300 K. Therefore, there is a wide gap between the primary crystallization of γ-Fe and $Nd_2Fe_{14}B$ crystallization, as shown in figure 20.7(a). Amorphous formability is rather poor and

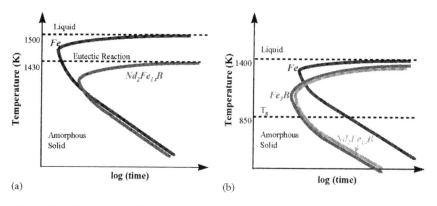

Figure 20.7. Concepts of TTT diagrams for (a) $Fe/Nd_2Fe_{14}B$ and (b) $Fe_3B/Nd_2Fe_{14}B$ nanocomposites.

type I processing is favourable. It is important to cool rapidly enough to prevent unfavourable grain growth of the Fe crystallites. On the other hand, Fe_3B-based systems are located near the bottom of a deep valley of eutectic solidification. The liquidus is pushed down and the kinetics of crystallization are slowed down, as shown in figure 20.7(b). Amorphous formability is large and type II processing is favourable, with a large processing window.

Detailed TTT diagram information for crystallization in Nd-Fe-B amorphous alloys is, however, very limited. The critical cooling rate for

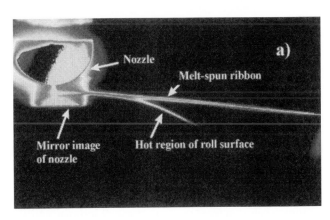

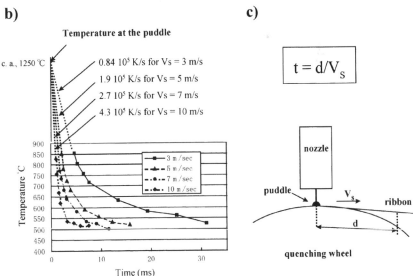

Figure 20.8. Example of measurement of cooling behaviour during melt-spinning process: (a) temperature mapping obtained from infrared imaging system; (b) cooling curves obtained from the temperature map for various surface velocities of the cooling wheel; and (c) and schematic geometry of the experiment.

amorphous solidification has been said to be around 10^5 to 10^6 K/s for typical Nd-Fe-B permanent magnetic alloys. It is only very recently that actual cooling rates during melt-spinning have been measured for a limited number of Nd-Fe-B nanocomposite alloys [6]. Figure 20.8 shows some examples of measurement of cooling curves for a $Fe_3B/Nd_2Fe_{14}B$-type nanocomposite alloy during the melt-spinning process. The temperature of the melt-spun ribbons was measured by utilizing an infrared imaging system similar to that first developed by [7]. Although the acquisition speed to capture an image is much slower than the travelling speed of the melt-spun ribbon, when the process is quasi-static in terms of temperature distribution, it is possible to obtain the cooling curve as a function of time after the melt is extracted from the puddle by using the simple relation among the travelling time, distance and velocity of the ribbon which is equal to the wheel surface velocity, as shown in figure 20.8(c).

This kind of observations enables a rough estimation of the nose position of the transformation curve along the time axis, although the information is limited to the case of CCT. Experimental determination of the temperature of the nose position is difficult. The lower arm of the TTT diagram for the time scale between a few tens of seconds and some hundreds of seconds may be obtained from isothermal measurements of the incubation time for crystallization. Such measurements are possible isothermally using a sensitive differential scanning calorimeter (DSC). Figure 20.9 shows a TTT diagram for

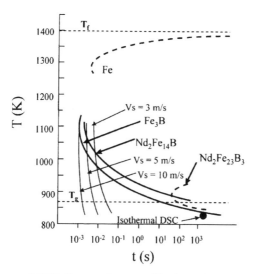

Figure 20.9. Concept of TTT diagram for crystallization of amorphous $Nd_4Fe_{77.5}B_{18.5}$ alloy. Thin solid lines represent cooling curves for different wheel velocities during melt-spinning process. The TTT curve for $Nd_2Fe_{23}B_3$ (broken curve) is added so that it may represent results of isothermal annealing experiments [6].

the $Nd_4Fe_{77.5}B_{18.5}$ system, which is a typical composition for an $Fe_3B/Nd_2Fe_{14}B$ nanocomposite. The diagram, however, is still only a rough concept relying upon limited number of independent observations, and may be subject to revision in future.

Micro-alloyed nanocomposite Nd-Fe-B magnets

One of the major aspects of designing the composition of nanocomposite materials is to manipulate the nanocrystalline structure, modifying the formation kinetics by alloying. Alloying with a small amount of IVB–VIB elements such as Cr, Nb and Zr has a profound impact on the solidification and crystallization kinetics of the Nd-Fe-B alloys. For an Fe-based nanocomposite, grain growth of primary Fe needs to be slowed down during rapid solidification (type I processing). On the other hand, for an Fe_3B-based nanocomposite which is produced by type II processing, the nucleation and growth rates of Fe_3B need to be optimized. Table 20.3 shows examples of additive elements and their influence on the solidification and crystallization behaviour of Nd-Fe-B nanocomposite alloys. Elements which have significant effects on the solidification and/or crystallization kinetics seem to be those which redistribute during the phase transformation. In Fe_3B-based nanocomposites, there are many metastable phases which compete kinetically and make crystallization very complicated.

α-Fe/$Nd_2Fe_{14}B$

α-Fe/$Nd_2Fe_{14}B$-based nanocomposites are expected to have a high saturation magnetization and therefore the potential to exceed single-phase $Nd_2Fe_{14}B$ magnets. In contrast to $Fe_3B/Nd_2Fe_{14}B$ nanocomposites, the amorphous formability of the Fe-rich alloys for the α-Fe/$Nd_2Fe_{14}B$-based nanocomposite is relatively poor. Fast quenching in the initial stage of rapid solidification is essential to avoid the development of coarse γ-Fe dendrites. In most cases, good α-Fe/$Nd_2Fe_{14}B$ nanocomposite magnets are obtained directly from the melt within a very narrow range of cooling rate. Addition of refractory metals such as Ti, V, Cr, Nb, Hf, Mo and W slows down crystalline growth of both Fe and $Nd_2Fe_{14}B$ particles during the rapid solidification process, as discussed below. This eventually widens the window of appropriate cooling rate. Nb is one of the most effective elements.

Modification of the growth kinetics during rapid solidification by adding TiC to Nd-Fe-B alloys has been applied successfully to develop alloy compositions suitable for the inert gas atomization process [8]. Kramer *et al* [9] have discussed the effect of TiC addition on the solidification process in $Nd_2Fe_{14}B$-based alloys in terms of the velocity of the solidification front and the temperature relative to the peritectic temperature of $Nd_2Fe_{14}B$

Table 20.3. Examples of additive elements and their influence on crystallization kinetics in Nd-Fe-B nanocomposite magnets.

Element	Influence on crystallization/solidification	
	α-Fe/Nd$_2$Fe$_{14}$B	Fe$_3$B/Nd$_2$Fe$_{14}$B
Ti	Retardation of growth rate of γ-Fe in melt [1]. Suppression of Nd$_2$Fe$_{23}$B$_3$ in B-rich alloys [1]	Suppression of Nd$_2$Fe$_{23}$B$_3$ [1]
V	Suppression of Nd$_2$Fe$_{23}$B$_3$ in B-rich alloys [1]	Suppression of Nd$_2$Fe$_{23}$B$_3$ [1]
Cr	Suppression of Nd$_2$Fe$_{23}$B$_3$ in B-rich alloys [1]	Partitioning in Fe$_3$B and shift of upper Nd concentration limit for Fe$_3$B/Nd$_2$Fe$_{14}$B formation [14]
Cu	Formation of Cu clusters but not acting as grain size refiner	Formation of Cu-rich region and Nd- and-B-rich region, and nucleation of Fe$_3$B there at low temperatures [16]
Ga		Enrichment of Ga in grain boundaries [20]
Zr	Grain size refinement	Retardation of Nd$_2$Fe$_{14}$B crystallization. Grain size refinement [22]
Nb	Grain size refinement [9]	Rejection of Nb from Nd$_2$Fe$_{14}$B. Grains size refinement and ormation of Fe$_{26}$B$_3$ [16]
Hf	Retardation of growth rate of γ-Fe in melt in B-rich alloy Suppression of Nd$_2$Fe$_{23}$B$_3$ in B-rich alloys [1]	Suppression of Nd$_2$Fe$_{23}$B$_3$ [1]
Mo	Retardation of growth rate of γ-Fe in melt in B-rich alloy	

formation. According to their model, the growth rate of Nd$_2$Fe$_{14}$B is significantly reduced by TiC addition, suppressing the growth transition from frequent (high density) nucleation of Nd$_2$Fe$_{14}$B in liquid supercooled below the peritectic temperature to dendritic growth of Fe in liquid heated up above the peritectic temperature by recalescence. Dendritic Fe is replaced by peritectic Nd$_2$Fe$_{14}$B when the liquid temperature cools rapidly, forming regions with coarse Nd$_2$Fe$_{14}$B grains. The growth kinetics in undercooled Nd$_2$Fe$_{14}$B alloys with C and Ti or Mo additions have recently been studied by Hermann and Bacher [10] using an electromagnetic levitation technique. The growth velocity of the Nd$_2$Fe$_{14}$B phase is estimated to be between 1.1 and 6.4 mm/s, depending on the degree of supercooling, in an Nd$_2$Fe$_{14}$B melt, but is reduced to 0.3–2.5 mm/s in an (Nd$_2$Fe$_{14}$B)$_{0.94}$(TiC)$_{0.03}$

alloy. The considerable slowing down of the growth kinetics results in refinement of the grain size.

The effects of Cr, Ti, Nb, Zr, Hf, Ta and W on the microstructure and magnetic properties of nanocomposites composed mainly of α-Fe and $Nd_2Fe_{14}B$ phase with a minor amount of ferromagnetic boride such as $Nd_2Fe_{23}B_3$ have been studied by Chang *et al* [11]. The formation of metastable $Nd_2Fe_{23}B_3$ is found to be suppressed by addition of Cr, Ti, Nb and V in $(Nd_{0.95}La_{0.05})_{9.5}Fe_{78}M_2B_{10.5}$. Thermal magnetic analysis indicates the existence of α-Fe and a $Nd_2Fe_{14}B$-type phase. Considerable refinement of grain size can be seen by TEM in $(Nd_{0.95}La_{0.05})_{9.5}Fe_{78}Cr_2B_{10.5}$ and in $(Nd_{0.95}La_{0.05})_{9.5}Fe_{78}Ti_2B_{10.5}$. In a similar composition range, Chiriac *et al* [12] report relatively good hard magnetic properties by chill disk melt spinning with a low surface velocity of only $3\,m/s$ to make $Nd_8Fe_{73}Co_5Hf_2B_{12}$. Thermal magnetic analysis indicates the formation of $Nd_2Fe_{14}B$ and α-(Fe,Co).

$Fe_3B/Nd_2Fe_{14}B$

The major shortcoming of $Fe_3B/Nd_2Fe_{14}B$ nanocomposites is their low coercivity. In order to obtain a higher coercivity, there are basically two factors to be improved.

1. The fraction of the hard magnetic phase, $Nd_2Fe_{14}B$ [4].
2. The magnetocrystalline anisotropy K_1 of the hard magnetic phase, $Nd_2Fe_{14}B$, by alloying with heavy rare earth elements such as Tb and Dy. However, grain size must be suitably refined in order to maintain the ratio between exchange coupling and magnetocrystalline anisotropy energy in the adequate range given by theoretical calculation [5].

Among various elements which have been tried to improve magnetic properties of the $Fe_3B/Nd_2Fe_{14}B$-based nanocomposites, chromium is the most important to enhance the intrinsic coercivity, H_{cJ} [13]. Because of the strong affinity of Cr with B, Cr is enriched in Fe_3B upon crystallization and stabilises this phase [14]. This leads to formation of the $Fe_3B/Nd_2Fe_{14}B$ composite even in the concentration range where formation of $Nd_2Fe_{23}B_3$ prevails, i.e. for Nd greater than about $5\,at\%$ in the ternary alloys. Accordingly, Cr addition helps to realize nanocomposites with a larger volume fraction of $Nd_2Fe_{14}B$ and hence a large coercivity. Suzuki *et al* [15] pointed out that Cr has a significant effect on the kinetics of phase formation and decomposition, and that it allows a reaction path in which the $Fe_3B/Nd_2Fe_{14}B$ composite is formed as a metastable intermediate structure instead of the $Fe_3B/Nd_2Fe_{23}B_3$ combination.

Niobium, on the other hand, stabilizes $Fe_{23}B_6$, and retards the decomposition of the $Fe_3B/Nd_2Fe_{14}B$ composites. $Fe_{23}B_6$ has a spontaneous magnetization of approximately $1.7\,T$ at room temperature [2], which is larger than that of Fe_3B (1.6 T) and $Nd_2Fe_{14}B$ (1.6 T). Therefore, the presence of this phase may be beneficial. According to Ping *et al* [16], $Fe_{23}B_6$ crystallizes from the residual amorphous phase nearly simultaneously with $Nd_2Fe_{14}B$ slightly above the crystallization temperature of Fe_3B. The resultant microstructure is characterized by finely divided crystalline phases between Fe_3B crystallites.

Copper has also a prominent effect on the kinetics of Fe_3B crystallization. Three dimensional atom probe microanalysis (3D-APM) has revealed that Cu-enriched clusters ($10^{24}\,m^{-3}$) are formed upon annealing of an amorphous $Nd_{4.5}Fe_{76.8}B_{18.5}Cu_{0.2}$ alloy well below the crystallization temperature of Fe_3B [16]. Nd is also enriched in these clusters, and they are surrounded by Nd-depleted and B-enriched regions which provide chemically favoured crystallization sites for Fe_3B [17].

Compositional adjustment has yielded a relatively high-coercivity material on the basis of Nd-Fe-B-Cr-Co, namely $Nd_{4.5}Fe_{73}B_{18.5}Cr_2Co_2$ which has B_r of 1.05 T, H_{cJ} of 378 kA/m and $(BH)_{max}$ of 108 kJ/m^3. Simultaneous addition of small amounts of Cu and Nb to this material has been successfully carried out to improve $(BH)_{max}$ with a slight sacrifice in H_{cJ} to yield $Nd_{4.5}Fe_{72.3}B_{18.5}Cr_2Co_2Cu_{0.2}Nb_{0.5}$ with $B_r = 1.10$ T, $H_{cJ} = 336$ kA/m and $(BH)_{max} = 123$ kJ/m^3.

Finally, gallium is quite important in improving magnetic properties of the $Fe_3B/Nd_2Fe_{14}B$ nanocomposites [18]. One of the best $Fe_3B/Nd_2Fe_{14}B$-type nanocomposites is $Nd_{3.5}Dy_1Fe_{73}Co_3Ga_1B_{18.5}$ with $(BH)_{max} = 136$ kJ/m^3, $B_r = 1.18$ T and $H_{cJ} = 390$ kA/m [19]. Mishra and Panchanathan [20] have studied $Nd_{3.5}Dy_1Fe_{73}Co_3Ga_1B_{18.5}$ by transmission electron microscopy using energy dispersive x-ray microanalysis (TEM-EDX), and observed that Ga additives are all detected in the $Nd_2Fe_{14}B$ phase with Ga concentrations mostly near grain boundaries. More recently, a 3D-APM analysis has revealed that Co and Ga atoms are partitioned to the $Nd_2Fe_{14}B$ phase, and evidence for a slight enrichment of Ga atoms at the $Nd_2Fe_{14}B/Fe_3B$ interface has been found [21].

To utilize heavy rare earth elements, manipulation of the crystallization behaviour, especially $Nd_2Fe_{14}B$, can be carried out by readjusting the Cu content so that the effect on the crystallization temperature of $Nd_2Fe_{14}B$ counteracts that of the heavy rare earth element. An example of such compositional modifications to yield higher coercivity magnets is $Nd_{3.5}Dy_1Fe_{71.3}B_{18.5}Cr_{2.4}Co_{2.4}Cu_{0.4}Nb_{0.5}$ with $B_r = 0.93$ T, $H_{cJ} = 468$ kA/m and $(BH)_{max} = 100$ kJ/m^3 [22]. Utilizing the stronger effect of Zr to prevent grain growth, better magnetic properties have been obtained in $Nd_{3.4}Dy_1Fe_{71.7}B_{18.5}Cr_{2.4}Co_{2.4}Cu_{0.4}Zr_{0.2}$ with $B_r = 0.97$ T, $H_{cJ} = 465$ kA/m and $(BH)_{max} = 105$ kJ/m^3. Figure 20.10 shows the coercivity versus remanence map as a summary of the compositional investigation based on the information of micro-alloying effects on the crystallization behaviour.

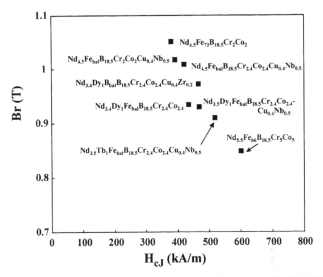

Figure 20.10. Compositions and magnetic properties of advanced $Fe_3B/Nd_2Fe_{14}B$-type nanocomposite permanent magnets.

Applications

Table 20.4 points out some of the unique aspects, features and application areas of Nd-Fe-B nanocomposite permanent magnets:

1. Because of the presence of the non-rare-earth phases such as Fe, the total rare-earth content is much less than conventional Nd-Fe-B magnets. As a result, the nanocomposites are much more stable against oxidation. The powders are easy to handle in a high-temperature process like compounding and inserting into molten plastic for injection moulding.

Table 20.4. Application candidates for isotropic NCPM powder.

Unique aspects	Features	Applications
Low rare-earth composition	Stable against oxidation	Injection-moulded bonded magnets
Nanocrystalline structure	Uniformity of magnetic properties over wide range of particle size. Uniformity of B_{pp} in narrow-pole magnetizing	Stepping motors, magnetic scales
Composite structure	Large recoil of magnetization	Short range latch magnets

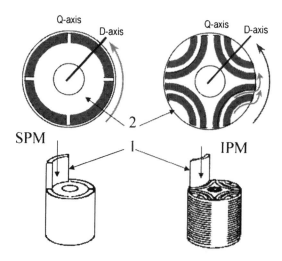

Figure 20.11. Rotor configuration of typical magnet motors. Parts 1 and 2 are, respectively, permanent magnet and laminated steel core [23]. Courtesy of Dr F Yamashita.

2. The nanocrystalline structure guarantees good homogeneity of magnetic properties over a wide range of particle size. This gives a uniformity of peak-to-peak magnetic flux distribution in the case of narrow-gap multi-pole magnetization.
3. The materials are composites containing a large volume fraction of a soft-magnetic phase. Consequently, the magnets show a relatively large recoil permeability, which Kneller and Hawig called an 'exchange-spring'.

An interesting utilization of the injection-moulding technique in the fabrication of permanent magnet motors has been recently proposed [23], in a study of two rotor configurations suitable for small d.c. motors with mechanical outputs of 1 W to 100 W, as shown in figure 20.11. The inner permanent magnet (IPM) rotor is found to be more efficient than the surface permanent magnet (SPM) rotor because the inner permanent magnet utilizes the reluctance torque associated with the magnetic flux path in the rotor. An $Fe_3B/Nd_2Fe_{14}B$-type nanocomposite is suitable for injection-moulding a thermoplastic compound into the curved gaps which serve as the permanent magnet layers after magnetizing. The morphology of $Fe_3B/Nd_2Fe_{14}B$-type nanocomposite powder particles, rounder than the thin flake-like morphology of conventional melt-spun Nd-Fe-B, is beneficial in providing a low viscosity compound for the injection process. The morphology of $Fe_3B/Nd_2Fe_{14}B$-type nanocomposite is closely related to the fact that the preferred roll velocity is much slower than the conventional Nd-Fe-B during rapid solidification, because of the high amorphous formability of the alloy.

Application of Nd-lean magnets with relatively small H_{cJ} values for multi-pole rotors for small stepping motors has also been proposed. The obvious difficulty is to maintain thermal stability in a practically tolerable range. The dependence of irreversible flux losses on the H_{cJ} values of several Nd-Fe-B-based nanocomposites has been studied by Yamashita *et al* [24]. At least 400 kA/m coercivity seems to be required to suppress irreversible losses in motors.

Apart from motors, various other applications may be possible for Nd-Fe-B nanocomposite permanent magnets making use of their excellent corrosion resistance, high remanence and potentially low material cost. The application area may include hybridizing this class of hard magnetic powder with other powders of conventional Nd-Fe-B or hard ferrite. The low rare-earth content is of major advantage in this class of material especially when a large amount of usage is expected, for instance in automobile applications.

Textured nanocomposites

Nanocomposite permanent magnets established to date are isotropic, with magnetic properties not superior to existing commercial products. An inescapable effect of introducing a large amount of a soft magnetic component is the reduction of the intrinsic coercivity. In addition, in a randomly oriented uniaxial ferromagnet the magnetocrystalline anisotropy is averaged out over a space covered by the exchange length and is consequently reduced as a result of strong correlation of magnetization direction due to exchange coupling amongst the grains. To avoid this effect, the crystallographic orientations of the hard magnetic phase must be aligned.

The above discussion merges with theoretical predictions of a giant energy product in textured nanocomposite magnets [3]. However, realization of a textured nanocomposite in which component phases couple with each other via exchange interactions is by no means easy. The strong tendency of thin films to develop texture has been the only successful way to fabricate an anisotropic nanocomposite. For instance, Parhofer *et al* [25] prepared a sandwich $SiO_2/Nd_2Fe_{14}B/\alpha$-$Fe/Nd_2Fe_{14}B/Cr$ film by sputter deposition with $B_r = 1.2$ T and $H_{cJ} = 380$ kA/m.

A good way to generate texture in bulk nanocomposite permanent magnets has not been discovered so far. Conventional processes depend on long-range diffusion and cannot produce the required nanometre grain sizes. Texturing by die-upsetting is effective for rapidly solidified and hot-pressed Nd-Fe-B with excess Nd [26], but has not been reported for hot-pressed Nd-Fe-B nanocomposites. Solution–precipitation creep produces material transfer via the grain boundary Nd-rich liquid phase during hot-deformation of fine crystalline Nd-Fe-B alloys, but cannot apply in

α-Fe/$RE_2Fe_{14}B$ nanocomposites because there is no grain boundary liquid phase.

The short term future objective is to realize commercialization of state-of-the-art nanocomposite permanent magnets, using their demonstrated advantageous features over conventional materials. The longer term objective is somehow to realize anisotropic nanocomposite magnets with clearly superior magnetic properties. A completely novel process must be developed in order to generate texture in bulk nanocomposite magnets. This is one of most challenging subjects for researchers in this field.

References

[1] Coehoorn R, de Mooij D B, Duchateau J P W B and Buschow K H J 1988 *J. de Phys. Colloque C8* **49** 669
[2] Kneller E F and Hawig R 1991 *IEEE Trans. Magn.* **27** 3588
[3] Skomski R and Coey J M D 1993 *Phys. Rev. B* **48** 15812
[4] Schrefl T, Fidler J and Kronmuller H 1994 *Phys. Rev. B* **49** 6100
[5] Fukunaga H, Kitajima N and Kanai Y 1996 *Mater. Trans. JIM* **37** 864
[6] Hirosawa S, Kanekiyo H, Shigemoto Y, Murakami K, Miyoshi T and Shioya Y 2002 unpublished research
[7] Gillen A G and Cantor B 1985 *Acta Metall.* **33** 1813
[8] Branagan, D J, Hyde T A, Sellers C H and Lewis L H 1996 *IEEE Trans. Magn.* **32** 5097
[9] Kramer M J, Li C P, Dennis K W, McCallum R W, Sellers C H, Branagan D J and Shield J E 1997 *J. Appl. Phys.* **81** 4459
[10] Hermann R and Bacher I 2000 *J. Magn. Magn. Mater.* **213** 82
[11] Chang W C, Wang S H, Chang S J, Tsai M Y and Ma B M 1999 *IEEE Trans. Magn.* **35** 3265
[12] Chiriac M, Marinescu M and Castano F J 2000 *J. Appl. Phys.* **87** 5338
[13] Hirosawa S and Kanekiyo H 1996 *Mater. Sci. Eng. A* **217–218** 367
[14] Sano N, Tomida T, Hirosawa S, Uehara M and Kanekiyo H 1988 *Mater. Sci. Eng. A* **250** 146
[15] Suzuki K, Cadogan J M, Uehara M, Hirosawa S and Kanekiyo H 1999 *J. Appl. Phys.* **85** 5914
[16] Ping D H, Hono K, Kanekiyo H and Hirosawa S 1999a *J. Magn. Soc. Japan,* **23** 1101
[17] Ping D H, Hono K, Kanekiyo H and Hirosawa S 1999b *Acta Mater.* **47** 4641
[18] Kanekiyo H, Uehara M and Hirosawa S 1993 *IEEE Trans. Magn.* **29** 2863
[19] Hirosawa S and Kanekiyo H 1994 in *Advanced Materials 93* ed. M Honma; *Trans. Met. Soc. Japan* **14B** 969
[20] Mishra R K and Panchanathan V 1994 *J. Appl. Phys.* **75** 6652
[21] Ping D H, Hono K and Hirosawa S 1998 *J. Appl. Phys.* **83** 7769
[22] Miyoshi T, Kanekiyo H and Hirosawa S 2000 *Proc. 16th International Workshop on Rare Earth Magnets and Their Applications* (Sendai: Japan Institute of Metals) p 495

[23] Yamashita F and Yamagata Y 1999 *J. Magn. Soc. Jpn.* **23** 1117
[24] Yamashita F, Ohara F K, Yamagata Y and Fukunaga H 2000 *J. Magn. Soc. Jpn.* **24** 431
[25] Parhofer S M, Wecker J, Kuhrt C, Gieres G and Schultz L 1996 *IEEE Trans. Magn.* **32** 4437
[26] Lee R W, Brewer E G, Schaffel N A 1985 *IEEE Trans. Magn.* **21** 1958

Index